ENVIRONMENTAL CAUSES AND PREVENTION MEASURES FOR ALZHEIMER'S DISEASE

ENVIRONMENTAL CAUSES AND PREVENTION MEASURES FOR ALZHEIMER'S DISEASE

GEORGE J. BREWER

ELSEVIER

ACADEMIC PRESS

An imprint of Elsevier

Library of Congress Cataloging-in-Publication Data
A catalog record for this book is available from the Library of Congress

British Library Cataloguing-in-Publication Data
A catalogue record for this book is available from the British Library

ISBN: 978-0-12-811162-8

For information on all Academic Press publications visit our website at
https://www.elsevier.com/books-and-journals

Working together
to grow libraries in
developing countries
www.elsevier.com • www.bookaid.org

Publisher: Nikki Levy
Acquisition Editor: Natalie Farra
Editorial Project Manager: Kathy Padilla
Production Project Manager: Stalin Viswanathan
Designer: Miles Hitchen

Typeset by TNQ Books and Journals

I dedicate this book to my wonderful wife, Lucia (Feitler) Brewer. She has devoted herself to me through good times and bad times, always loyal, always supportive, and always ready to laugh at my sometimes silly jokes. She started out as a research assistant in my laboratory, and the rest is history. She always has good advice and I received plenty of it in writing this book. We just celebrated our 47th wedding anniversary and I am ready for 47 more. I love her and love living with her, and I appreciate all she does for me.

CONTENTS

FOREWORD

In less than 40 years, Alzheimer's disease (AD) has transformed from unknown to the most dreaded ailment existing today. Nearly half of those over 85 years of age are affected. Is the increase from awareness, incidence, or both? We do not know for sure, but Dr. Brewer presents evidence that increased AD may result from ingestion of drinking water containing divalent copper-2 rather than the monovalent copper-1 found in food. In contrast to copper-1, which is rapidly stored in the liver, copper-2 is also accumulated in the brain.

Accumulated copper may play a critical role in the increased AD seen in developed societies or the several-fold differential in prevalence seen between ethnic groups. Dr. Brewer makes several persuasive arguments, among them that copper's strong interaction with amyloid β protein precursor and amyloid β reduces redox activity. Dr. Brewer further provides critical public health implications of copper dysregulation and the need for further study. His synthesis offers hope to AD research for prevention and treatment.

George Perry, PhD
College of Sciences
The University of Texas at San Antonio
San Antonio, Texas

ACKNOWLEDGMENTS

First I would like to acknowledge the invaluable help of my chief word processor, Ann Arnold. Ann has the patience of a saint, and she needed it to work through and type my long hand scribbling as I wrote this book. She also helped keep this relatively complex project organized. Thanks also to my loyal former secretary, Julia Sitterly, who still helps me with some word processing, and keeps and sends out as necessary personal office items such as my CV, my picture, etc. I also appreciate the assistance of my loyal and excellent former chief technician, Robert Dick, who still does some moon-light work in our laboratory, formats data, trouble shoots my computer, and helps me find references, including many for this book.

I also would like to thank the excellent staff at Elsevier, in particular Natalie Farra, and Kathy Padilla, both of whom were extremely helpful in the publishing process.

CHAPTER ONE

Introduction

He who finds a thought that enables him to obtain a slightly deeper glimpse into the eternal secrets of nature has been given great grace.

Albert Einstein

As stated above so well by Einstein, we scientists feel a touch of awe, or grace, when we uncover a new scientific truth. I now feel that sense of grace after uncovering a major environmental cause of Alzheimer's disease (AD). AD is at a catastrophically increasing frequency and causes a disastrous loss of ability in our elderly to think and act effectively, thus robbing them of their golden years. Since the cause I have discovered can be easily corrected, or avoided, by individual persons, the grace I feel will be multiplied several fold if I can get this message out to the academic community and later the general public, and if it results in a major curbing of the disaster that is AD. So the purposes of this book are (1) to educate academicians and others about this newly discovered environmental cause, providing enough scientific evidence to convince, but not so much as to confuse and (2) to show people how to make the changes that will make all the difference in preventing AD.

In this introductory chapter, I will outline rather broadly what will be covered in the rest of the book. By the end of this chapter, you the reader will have the big picture, but not the details behind this major cause of AD. The details will be filled in as you read the rest of the chapters.

First of all, what is AD? In Chapter 2, we will provide more details about the disease that has reached epidemic proportions in many parts of the world, including the United States, but here we will just say enough to make sure everyone has a basic concept of the disease. It usually occurs in people aged 60 years and older, and usually begins with memory loss. If only mild memory loss is affected, it is called mild cognitive impairment (MCI). But MCI is usually a precursor to AD, as 80% of MCI patients convert to full AD at a rate of 15% per year. As the disease progresses, other cognitive functions in addition to memory begin to be lost. Decision making deteriorates, and the ability to perform even routine tasks is lost. Eventually the patient requires a full-time caretaker. Very often this is a family member, such as a spouse, so that the disease

Environmental Causes and Prevention Measures for Alzheimer's Disease
ISBN 978-0-12-811162-8
http://dx.doi.org/10.1016/B978-0-12-811162-8.00001-9

results not only in loss of enjoyment of the golden years of the patient but also puts a tremendous burden on the spouse or other members of the family, or whoever is the caregiver. This is particularly burdensome, because the AD patient may live for a decade or two requiring a full-time caretaker.

Memory and the ability to do tasks appropriately are all part of what is called cognition, so the AD patient loses more and more cognition as the disease progresses. But what is going on in the brain of the AD patient to cause this loss of cognition? Are there abnormalities that can be seen in the brain? Yes, scans of various types will show shrinkage of the brain. And at autopsy, typical abnormalities are seen in the brains of AD patients. Besides shrinkage, a prominent abnormality that is seen is what are called "amyloid plaques." A piece from a protein called amyloid precursor protein is clipped off. This piece is called beta amyloid, and large numbers of beta amyloid pieces clump together to form amyloid plaques. These plaques are uniquely present in AD brains, and so characteristic that many experts think they play a critical role in the disease process, causing damage to the brain. Found bound to these plaques are large quantities of three metals, zinc, copper, and iron. The binding of copper and iron in these plaques causes them to generate what are called "oxidant damaging radicals" in the brain. This oxidant damage does appear to be involved in the injury to the AD brain. More details are given about amyloid plaques in Chapter 2 and the metals bound to them in Chapter 4.

A second prominent abnormality found in AD brains at autopsy is called "neurofibrillary tangles." These are tangles of the hairlike outgrowths from the neurons (brain cells) of the brain. Amyloid plaques and neurofibrillary tangles are so specific for AD that it allows the pathologist to make a certain diagnosis of AD. In fact, this is the only way the diagnosis of AD can be made with 100% certainty. However, doctors who have expertise with AD patients can nowadays make the diagnosis in live patients with about 95% accuracy.

In Chapter 3 we will provide interesting new facts about the epidemic of AD that most people, and even doctors who treat lots of AD patients, don't realize. These facts provide a whole new perspective about AD. These facts shout out, "AD is caused by something new in our environment!" This new perspective is both depressing and exciting. Depressing because it turns out the epidemic, and the loss of their golden years by our elderly, is mostly caused by something we are doing to ourselves. Exciting because there are some simple things we can do to reverse the epidemic.

In a nutshell, these interesting new facts are

1. AD is at epidemic proportions in developed but not undeveloped countries.

2. AD at a high frequency is a relatively new disease phenomenon, reaching epidemic levels in the last 100 years. AD was nonexistent, or very rare, before the year 1900. The evidence for this is very good.

3. The absence of AD in the 1800s is not due to the lack of elderly people in high-enough frequency to have been affected by the disease.

4. The absence of AD in the 1800s is also not due to it not being noticed, or attributed to "normal aging."

This set of facts lead to a clear conclusion, almost "shout it out" as said earlier, that some thing or things in the new environment in developed countries, not present in undeveloped countries, is, or are, strongly causal of the AD epidemic.

I would like to emphasize to readers how important the above conclusion is. It is more important than finding out that cigarette smoking causes cancer, heart disease, and stroke. At least these killers often kill relatively fast, and generally patients retain their thinking ability until the end. That is, they remain human until the end. Not so with AD. Patients are gradually stripped of their memories and thinking ability over a period that may last up to two decades. They lose what makes them human and become no more than pet animals. Is that the way we want our loved ones to end up their lives, with a caretaker caring for them for years and years like a pet animal?

So surely, if a major cause of the disease is something in the environment, we as a society and as individuals should leave no stone unturned to find the cause and remove it, so as to prevent as many cases of this degrading disease as possible. Well, we now know a major environmental cause. As we will discuss in Chapters 6–9 in detail, it is ingestion of inorganic copper, together with a high meat diet. In Chapter 11, we tell you how to eliminate these risks, which can be done in a relatively simple, inexpensive manner. This message needs to be disseminated in our society, so that people can learn how to avoid this very bad disease.

Continuing with the broad outline of the book, having presented in Chapter 3 the clear evidence leading to the conclusion that a new environmental agent or agents is causing the AD epidemic, in the next two chapters, Chapters 4 and 5, a logical search of what those agents might be is pursued. To be identified as a likely causative environmental factor, the factor (or agent) must pass two tests. First, does change in the factor or agent fit with the AD history and demographics demonstrated in Chapter 3. That is, is it now different between developed and undeveloped countries, and has it changed in developed countries over the last 100 years? The second test is the question of whether there are data of some type making the factor or agent a logical causative factor for AD. It must be recognized that in terms of this second test, not everything is currently known about all the ways AD can be caused. So just because a factor or agent fails this second test doesn't rule it out as a possible candidate. It's just that the case is greatly strengthened if the second test is positive as well as the first. In Chapter 4, the search will focus on metals. Several metals have been suspected of playing AD causative roles, and if exposure has changed, one or more of them could be a causative environmental factor. We will come out of this search by hitting pay dirt. Copper turns out to be the culprit, and in two ways. Exposure to copper-2 (divalent copper) is specifically AD causative, and a lifetime exposure to increased body copper levels is also AD toxic. Copper-2 and copper, in general, both pass tests for being environmental AD causative agents. Much is being written in the recent literature about the effect of diet and other lifestyle factors on AD. So, in Chapter 5, whether any lifestyle factors could explain the AD epidemic is examined. In Chapter 5 we hit pay dirt again, identifying increased meat eating with increased copper absorption as a likely new environmental AD causative factor.

In Chapter 6, detail about how copper-2 and copper in general fits as major environmental factors causing the AD epidemic will be provided and how increased meat eating fits as well, because it increases copper absorption.

Continuing with the broad outline of the book, in Chapter 7 background on copper in general is provided, since much of the book will be talking about copper. Copper's role in the body as an essential element will be discussed. Unlike lead, which is purely a toxic metal, while copper can be toxic, for example as said here copper toxicity is a cause of AD, it is also critical to life, taking part in many vital reactions.

Also in Chapter 7, the difference between what is called "inorganic copper" and "organic copper" will be discussed. In brief, organic copper is the copper in food, safely bound to proteins and other organic molecules in the food. It is primarily copper-1. This organic copper is absorbed from the intestine and is taken immediately to the liver, where it is put in safe channels. Inorganic copper, which is copper-2, such as copper in drinking water or in mineral supplement pills, are simple salts of copper, not bound to anything. A portion of inorganic copper is absorbed directly into the blood, bypassing the liver, and directly contributing to the potentially toxic copper of the body. Think of it this way: Our bodies evolved to safely handle the copper in food. Our bodies have not evolved to safely handle inorganic copper, because in the prior eons of evolution, inorganic copper was not ingested. It is only during the 1900s and since that humans have been exposed to copper in drinking water, because of the spread of copper plumbing in developed countries, as well as the increasing habit in developed counties of taking a daily multimineral pill containing inorganic copper. The reason for the difference in the way the body handles inorganic copper and organic copper is now known. As explained by reviewing the simple chemistry of copper in Chapter 7, the two coppers are chemically a little different, due to a difference in valence. This little difference makes all the difference in whether copper is toxic or not. Think of inorganic copper as if it were lead! It is almost as poisonous as lead!

In Chapter 8, we will describe how we came to suspect that ingestion of inorganic copper, or copper-2, in drinking water was causal of AD, and the web of evidence supporting that contention. But here I'd like to recount the story of an accidental discovery that was key to realizing the exquisite toxicity of copper in drinking water.

So, here is the fascinating story about the serendipitous accidental discovery that sparked a whole new body of research. During the 1990s, Dr. Larry Sparks working in West Virginia became interested in the connection between cholesterol and AD. He had a job working as a pathologist at the county morgue where part of his duties involved examining tissues of autopsy subjects who died of unknown causes. When he examined brain sections from persons who were later confirmed to have died of a heart attack, he made the observation that all too often these heart attack victims also had the plaques and tangles in their brain slices that are the hallmark feature of AD. What he discovered was that these heart attack victims were on their way to developing AD, but they died of a heart attack first!

Based upon Dr. Sparks' discovery of a connection between cholesterol and AD, he began to think about ways that AD might be combated with the same type of drugs used to reduce the risk of heart attack. These are cholesterol-lowering drugs, such as the bestselling statin-based drugs: Lipitor, Zocor, and Mevacor, for example. He then went about the next logical research step, developing animal models of AD so that he could test both his hypothesis and the various cholesterol-lowering drugs for their effects on Alzheimer's-like disease in animals. While in West Virginia, he was able to create an animal model of AD in rabbits by feeding them a high-cholesterol diet for 10 weeks. In some months after the high-cholesterol feeding, these rabbits developed brain pathology (amyloid plaques) and deficits in learning and memory similar to that seen in human AD patients.

Keenly interested in his research and the potential to extend the use of these drugs into diseases such as AD, Dr. Sparks was courted by and began collaborating with various drug companies, including Parke-Davis, now a subsidiary of Pfizer, and maker of the best-selling statin, Lipitor, with over $13 billion in annual sales. His pioneering discoveries, notoriety, and animal models exploring the relationship between cholesterol and AD also caught the attention of research institutions having an interest in solving the riddle and epidemic that is AD. One of these institutions, Sun Health Research Center in Phoenix, Arizona, having a very large elderly population and interest in AD research, recruited Dr. Sparks to join them where they offered him a new facility and resources to carry out his work. Dr. Sparks took the offer and set about transferring his research operation from West Virginia cross country to Phoenix, Arizona. Awaiting Dr. Sparks in Arizona was the finding that would change the Alzheimer debate forever!

Once established at his new facility at Sun Health Research Center in Phoenix, Dr. Sparks set about recreating his high-cholesterol rabbit model of AD, which had been well established back in West Virginia. Since these rabbits must be fed a high-cholesterol diet for several months, reestablishing this model in Arizona would take the good part of a year from initial planning to final analysis. Unfortunately, when he subjected rabbits at the new facility to his typical battery of tests to determine deficits in learning and memory, unlike his "West Virginia bunnies" these "Arizona bunnies" behaved completely normal! When he examined brain slices from the Arizona animals, they lacked the characteristic brain lesions found in the West Virginia animals. For a researcher trying to establish animal models of disease, getting healthy animals is the worst of outcomes, especially when you are under the scrutiny of a new institution that just made a substantial investment of time and money in your work.

What had gone wrong in Arizona? Dr. Sparks began to ponder the variables. The rabbits were the same breed and type, even obtained from the same breeder as he used in West Virginia. The procedures used to rear the rabbits were all the same. The Arizona rabbits were fed the same high-cholesterol chow obtained from the same manufacturer as the chow used in West Virginia. At the peak of his frustration, he assembled his staff around the animal cages with the rabbits looking on and went step-by-step through every procedure that they had conducted. Suddenly, Dr. Sparks looked up and saw two large blue plastic bottles hung on the wall which appeared out of place. Dr. Sparks put the question to his staff, "What are those blue bottles on the wall?" "Those blue bottles hold the distilled drinking water for the rabbits," they replied.

Whereupon, Dr. Sparks exclaimed, "I didn't give the rabbits distilled water in West Virginia, I gave them tap water": Tap water turned out to be the one key variable that had changed between West Virginia and Arizona.

Disheartened but determined, Dr. Sparks spent the next year recreating his high-cholesterol-fed rabbit models, except this time he separated the animals into various groups. While all the new rabbits would be fed a high-cholesterol diet, one group would drink distilled water for 10 weeks and another group would drink tap water for 10 weeks. To identify the component of tap water that might be responsible for the Alzheimer's-like disease, different groups of rabbits would drink distilled water separately supplemented with trace amounts of the minerals found in the tap water in West Virginia, such as copper, zinc and iron. After many additional months, low and behold, the group of high-cholesterol-fed rabbits that drank tap water and the group that drank distilled water with added trace amounts of copper developed the Alzheimer's-like disease. Having made his second important discovery in Alzheimer's research, in addition to the role of cholesterol, Dr. Sparks and his colleague at West Virginia University, Bernard Schreurs, published their results in 2003.[1] Their paper appeared in the prestigious peer-reviewed journal, Proceedings of the National Academy of Sciences, and their publication was entitled, "Trace amounts of copper in water induce beta-amyloid plaques and learning deficits in a rabbit model of Alzheimer's disease." The Alzheimer's-like disease not only produced increased amyloid accumulation and amyloid plaques but also dramatically affected the memory of the rabbit, as shown in Fig. 1.1.

Dr. Sparks' findings published in 2003 showing that trace amounts of copper in drinking water at concentrations of only 1/10th the current actionable limit set by the Environmental Protection Agency, shocked the international Alzheimer's research community. This community was not immediately willing to accept these findings from a researcher who at the time was still considered outside the mainstream of Alzheimer's research. This perceived reluctance on the part of the Alzheimer's research community may be attributed to a number of factors. Chief among them was a lack of information about copper among professionals who deal with Alzheimer's patients and Alzheimer's research, and a lack of information about copper and the copper threat among the public. If people are aware of a medical threat, they ask their doctors about it, which puts pressure on professionals to educate themselves about the threat. This book is designed to educate both the public and professionals about the copper threat, which I called the "Copper Hypothesis".

The paper by Sparks and Schreurs in 2003[1] led to an epiphany in my thinking about the cause of AD. It made me realize that copper in drinking

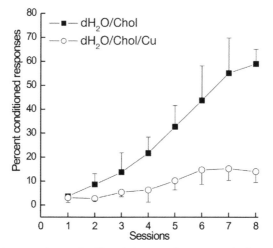

Figure 1.1 The large and very significant decrease in memory in the rabbits receiving trace amounts of copper (0.12 ppm) in their drinking water is shown. *From Sparks DL, Schreurs BG. Trace amounts of copper in water induce β-amyloid plaques and learning deficits in a rabbit model of Alzheimer's disease. Proc Natl Acad Sci 2003;100(19):11065–69.*

water was much more toxic to the brain than the copper in food. Very soon thereafter, the study by Morris and colleagues[2] that the copper in supplement pills caused a rapid loss in cognition led me to modify my hypothesis to the "Inorganic Copper Hypothesis." I have already explained what inorganic copper is, and because of its slightly different chemistry, it is absorbed differently, directly into the blood, compared to organic copper.

In Chapter 8, we will go into detail about the overwhelming web of evidence we have put together to show that ingestion of inorganic copper, or copper-2, is a major causal factor in AD. In Chapter 9, we will review the evidence that an increased body load of copper in general, irrespective of whether it is achieved by ingestion of organic or inorganic copper, is a risk factor for AD. Then, we will show why the body load of copper has increased over the last century, due to the increased ingestion of meat, from which copper is much better absorbed than from vegetable foods. Chapter 10 will discuss how nicely the copper causation mechanism fits with current overarching theories of AD causation, such as the amyloid cascade hypothesis.

In Chapter 11, we will tell you the two relatively simple steps that people can take to protect themselves against the toxicity of inorganic copper, and thus protect themselves against AD. Chapter 12 will discuss failures. The federal government through the Environmental Protection Agency (EPA) has failed to protect the population against copper in drinking water allowing over 10 times the level Sparks and Schreurs[1] showed caused AD-like disease in animal models. The original study was published 13 years ago and has been replicated several times, but the EPA has taken no action. While it is true in most regions that the source water provided to homes is quite low in copper, and the main culprit is copper plumbing in the home, in some regions source water as high in copper as the EPA allows could be a major factor in ingestion of inorganic copper and AD causation.

Then there is the failure of the Food and Drug Administration (FDA) to take action on the copper in multimineral supplement pills. It has been 10 years since Morris et al.[2] published that ingestion of these copper supplements, if the diet is also high in fat, increases cognition loss sixfold. Yet supplement pill makers continue to put 1–2 mg of copper in their pills, an extraordinarily large amount of inorganic copper, keeping in mind dietary food copper intake is only about 1 mg/day.

Chapter 13 discusses treatment of AD. Most of this book is about preventing AD, but the prevention measures discussed here may not be of much help to those who already have the disease. The FDA has approved five drugs for the treatment of AD, but all of them give only relatively small, and temporary, improvement. Behavioral symptoms may be a major problem in patients with AD, and a number of drugs, approved for other conditions, may benefit AD patients with behavioral problems. In brief, however, treatment is relatively ineffective, so it further drives home the message of this book—carry out the prevention measures before it is too late.

Chapter 14 provides a final summary of the book, taking the reader through the demographic and historical facts of the AD epidemic, which leads to the inevitable conclusion that the epidemic is caused by new environmental factors in developed countries in the 20th century. Then the search for these factors, leading to the conclusion that copper-2 ingestion from supplement pills and drinking water, together with an increased copper load from increased meat eating, is cause of the epidemic. More data supporting these conclusions are then provided. Prevention measures are covered in detail. Government failures to protect the citizens are discussed, and the relatively ineffective treatments developed so far.

So there you have the overview. Please read on to get the details of this critically important story.

REFERENCES

1. Sparks DL, Schreurs BG. Trace amounts of copper in water induce beta-amyloid plaques and learning deficits in a rabbit model of Alzheimer's disease. *Proc Natl Acad Sci USA* 2003;**100**:11065–9.
2. Morris MC, Evans DA, Tangney CC, et al. Dietary copper and high saturated and trans fat intakes associated with cognitive decline. *Arch Neurol* 2008;**63**:1085–8.

CHAPTER TWO

A Little Background on Dementia and Alzheimer's Disease

INTRODUCTION

Before delving into the main theme of this book, which is to show that Alzheimer's disease (AD) is mostly caused by ingestion of inorganic copper, or copper-2, in drinking water and in supplement pills containing copper, and will very likely be mostly preventable by avoiding ingestion of inorganic copper, a basic description of AD will be provided in this chapter so that everyone has a general factual knowledge of the disease. So here, dementia (loss of cognition) in general and then the dementia of AD and the overall disease process will be described.

DEMENTIA AND ITS CAUSES

Dementia refers to the loss of cognition, or thinking ability, and there are three main causes, vascular dementia, Lewy body dementia, and AD.

Vascular dementia is the second-most common cause of dementia, behind AD. It is due to damage to the brain from a loss of blood supply to a portion of the brain used for cognition. In general this is due to obstruction of one or more brain blood vessels by a blood clot. If a major brain blood vessel is obstructed, it is called a stroke, which is often manifest in part by a loss of control of muscles in some part of the body. But the most common events leading to vascular dementia are a series of small vessel obstructions, largely unnoticed, over a period of years, gradually leading to obvious dementia.

Atherosclerosis (called hardening of the arteries in lay terms) is the underlying process leading to these small strokes. In atherosclerosis, the smooth lining of various arteries is interrupted and replaced with an atherosclerotic plaque. This roughened area leads to small blood clots forming. These clots are at first attached to the plaque but can break off into the bloodstream. If the blood vessel is one of the vessels supplying the brain, these tiny clots can travel to the brain and obstruct a small blood vessel providing needed oxygen and nutrients to a small segment of the brain, causing the neurons

Environmental Causes and Prevention Measures for Alzheimer's Disease
ISBN 978-0-12-811162-8
http://dx.doi.org/10.1016/B978-0-12-811162-8.00002-0

(the brain cells) in that area to die. If this type of damage is repeated enough in areas of the brain where cognition is carried out, cognition is slowly lost, until the patient has lost enough to be classified as having dementia.

Atherosclerosis, the main cause of vascular dementia, has a set of risk factors that tend to make it occur and make it worse. The major risk factors are untreated or inadequately treated high blood pressure, untreated or inadequately treated high blood cholesterol and other blood fats, diabetes, particularly if poorly controlled, obesity, and smoking. Since atherosclerosis is a disease of aging, vascular dementia is a disease of aging.

Since these risk factors can be treated medically or by life-style changes, much vascular dementia can be successfully prevented by appropriate actions. A useful prevention strategy for people who have developed atherosclerosis and are at significant risk for strokes is to take an antiplatelet agent daily. The mechanism of vascular dementia involves formation of small blood clots on the atherosclerotic plaques as said earlier. Blood clotting requires small particles in the blood called platelets. The action of platelets to initiate blood clotting can be inhibited (slowed) by taking an antiplatelet drug. These include a small dose of aspirin (81 mg, or a baby aspirin) daily, or a daily dose of drugs whose trade names are Aggrenox or Plavix. These drugs do not prevent blood clotting if a major cut or traumatic event occurs, so they are generally safe.

Lewy body dementia is the third-most common cause of dementia. It is so-called because of the presence of small inclusions in neurons called Lewy bodies. Like AD and vascular dementia, Lewy body dementia is also a disease of aging. Early symptoms include visual hallucinations and difficulties with some movements, reminiscent of Parkinson's disease. Fluctuations in cognition often exist from day to day. It progresses to death in an average of 7 years, similar to AD. Autopsy studies suggest it accounts for about 10%–20% of dementia.

In addition to these three, there are several other rarer causes of dementia listed in the 2016 Facts and Figures of the Alzheimer's disease Association.[1]

Also, it needs to be pointed out that quite frequently more than one cause of dementia coexists. This is called mixed dementia. It is estimated that about a half of older people with dementia have pathologic evidence of more than one cause of dementia.[1] AD and vascular dementia are very often found together as a mixed dementia.

Using cognitive tests to be discussed later, it is not too difficult to make a diagnosis of dementia. But the harder part is to determine the cause. First, it should be realized that there are some patients who have symptoms of dementia and test positive for dementia who do not actually have dementia but have symptoms from some cause that mimics dementia. Some causes of dementia-like symptoms are depression, thyroid disease, side effects of some medications, alcohol abuse, certain vitamin deficiencies, etc.[1] It is estimated that 9% of what appears to be dementia is due to dementia-like symptoms from some reversible cause such as those listed. So it is important to rule out this possibility first.

The presence of risk factors for atherosclerosis and evidence of atherosclerosis make vascular dementia a likely possibility. But the likely presence of vascular dementia does not rule out AD, because in a significant proportion of patients, there is a mixed dementia due to both AD and vascular dementia.

Of course, as said in Chapter 1, there is now a major epidemic of AD in developed countries, such that it is now the most common cause of dementia in these countries, with about 60%–80% of patients with dementia having AD in either single or mixed form. So now AD will be discussed in more detail.

ALZHEIMER'S DISEASE
The Clinical Picture of AD and How the Diagnosis is Made

The risk period for nongenetic AD begins at about age 60. The first thing that occurs is a loss of memory. Since generally as people age they have little memory lapses (often called "senior moments"), such as difficulty recalling some word or a name, the more serious memory problems leading to AD have to be differentiated and identified. A big help in that regard is a test called the Standardized Mini-Mental State Examination,[2] or SMMSE. The test is reproduced in Table 2.1. The test is a series of 12 questions requiring memory and some other basic cognitive questions and is simple enough to administer in a doctor's office. A perfect score is 30 (some questions have multiple parts, all scored). The test is considered in the normal range at 26 or above, scores of 25 to 18 are considered mild cognitive impairment (MCI). Below that, is the dementia range.

If only memory is mildly to moderately affected, it is called MCI, as already said. MCI is a precursor to AD, with a larger proportion eventually converting to AD. As AD progresses, cognitive functions besides memory are affected. The patient may have difficulty putting multistep tasks together, such as cooking, getting dressed, going to the grocery store.

Table 2.1 The Standardized Mini-Mental State Examination

		Score	Maximum score
1.	(Allow 10s for each reply)		
	a. What year is this?		1
	b. What season is this?		1
	c. What month of the year is this?		1
	d. What is today's date? (answer can be off by 1 day)		1
	e. What day of the week is it?		1
2.	(Allow 10s for each reply)		
	a. What country are we in?		1
	b. What state/province/county are we in?		1
	c. What city/town are we in?		1
	d. What building are we now in?		1
	e. What floor are we on?		1
3.	"I am going to name three objects. After I have said all three names, I want you to repeat them. Remember what they are because I am going to ask you to name them again in a few minutes." (The words are said slowly at 1-s intervals). "Ball – Car – Man. Please repeat the three items for me." (Score 1 point for each correct answer)		3
4.	"Spell the word WORLD." (You may help the subject with the spelling.) "Now spell it backward please."		5
5.	"Now what were the three objects that I asked you to remember? (Score one point for each correct answer regardless of order, allow 10s)		3
6.	Show the subject your wristwatch or watch. "What is this called?" (Accept wristwatch or watch. Clock or time are not acceptable).		1
7.	Show the subject a pencil. Ask, "What is this called?" (Accept only pencil)		1
8.	"Please repeat this phrase after me: No ifs, ands, or buts." Answer must be completely accurate.		1
9.	Say, "Read the words on this piece of paper and then do what it says." Hand the subject a piece of paper with CLOSE YOUR EYES written on it; the only correct answer is if the subject closes their eyes. It does not matter if they read it aloud or not		1

Table 2.1 The Standardized Mini-Mental State Examination—cont'd

10.	Ask if the subject is right- or left-handed. If right-handed take a piece of paper, hold it up in front of subject and say, "Take this paper in your right hand, fold the paper in half with both hands and put the paper down on the floor."		
	Takes paper in correct hand		1
	Folds paper in half		1
	Puts paper on floor		1
11.	Hand the subject a pencil and a piece of paper and say, "Write any complete sentence on that piece of paper." Score 1 point if they write any sentence that makes sense. Ignore spelling.		1
12.	Place a piece of paper on the table. Make a simple design such as two interlocking pentagons on the paper. Give the subject a piece of paper and a pencil and say, "Please copy this design." Score one point if the design is essentially correct.		1
	Total Score	–	30

Modified from Molloy W, Caldwell P. Alzheimer's disease. Toronto: Key Porter Books; 1998. p. 76–78.

As these somewhat more complex functions begin to be interfered with, the doctor should consider the diagnosis of AD. But AD can only be definitively diagnosed at autopsy, where the brain reveals the diagnostic amyloid plaques and neurofibrillary tangles. So the physician faced with a patient with a dementia diagnosis does the next best thing, which is look at the probability of AD and the other two, somewhat common, causes of dementia, namely vascular dementia and Lewy body dementia. The physician reviews the history of the patient for the risk factors for AD, discussed in the next section, and vascular dementia and Lewy body dementia. The risk factors for vascular dementia were mentioned earlier, namely high blood pressure, high blood cholesterol, diabetes, obesity, and smoking. If these are not present, or have been effectively treated, and the patient has no clinical evidence or history of atherosclerosis, a diagnosis of vascular dementia is unlikely. However, it must be remembered, even if they are present and vascular dementia is a good possibility, AD may also be present in a mixed dementia. Lewy body dementia can usually be ruled out by the clinical picture. Today, doctors seeing large numbers of patients with dementia can make the diagnosis of AD in patients who are alive with about a 95% accuracy.

The Brain Pathology of Alzheimer's Disease

While the patient with AD is suffering memory loss and loss of other aspects of cognition, changes are occurring in the brain which, if the patient dies, can be seen at autopsy, under the microscope.

First, outside and around the neurons in parts of the brain responsible for cognition, there are dense collections of material. These turn out to be mostly aggregates of beta amyloid and are thus called amyloid plaques. Beta amyloid comes from a brain protein called amyloid precursor protein. An enzyme called beta secretase clips off the end of the amyloid precursor protein, and the resulting short piece of protein clipped off is the beta amyloid. This process is shown in Fig. 2.1.

It appears that this process of beta amyloid formation goes on in the normal brain, but the beta amyloid does not accumulate, aggregate, and form plaques as happens in the AD brain. Exactly how this process goes awry in the AD brain is not completely clear. Possibilities include an increased production or slowed removal of beta amyloid in the AD brain. Or a likely possibility is that something, possibly an increased copper level in the AD brain, causes the beta amyloid to aggregate into

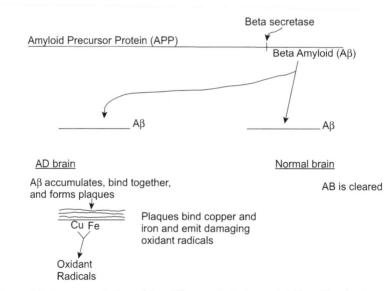

Figure 2.1 A representation of the difference in beta amyloid handling by the normal and the AD brain. Beta secretase cleaves a piece of APP off, and the piece is called beta amyloid (Aβ). In the normal brain, Aβ is cleared. In the AD brain, Aβ accumulates and binds together into amyloid plaques; these plaques can bind copper and iron, and this causes release of damaging oxidant radicals.

clumps before it can be removed. It is known that copper causes beta amyloid aggregation.

Copper or iron binding to the amyloid plaques causes formation of oxidant radicals, molecules that cause a type of damage called oxidant damage, which causes inflammation. This type of damage can cause injury to various molecules and lead to death of the neurons. It is known that the AD brain is characterized by excess oxidant damage and excess inflammation. It is also known that excessive death of neurons is occurring in AD. Cell death may in large part be caused by oxidant damage from copper and iron bound to amyloid plaques.

The amyloid plaques are so integral to the Alzheimer's disease process that most scientists working in this area believe they are causal of AD. This causal hypothesis is called the amyloid cascade hypothesis.[3] This hypothesis and another causal hypothesis will be discussed in detail in Chapter 10. The concept that amyloid plaques are causal has led to efforts to treat them in some manner. One approach is to use antibodies to beta amyloid, hoping to prevent accumulation and aggregation. So far, these efforts have been unsuccessful.

A second type of abnormality in the AD brain is tangles of fibers within neurons called neurofibrillary tangles. It is not clear how these form, but they undoubtedly represent some type of damage in these neurons, in other words, damage in the brain's key cells.

These two types of abnormalities, extracellular amyloid plaques and intracellular neurofibrillay tangles, are the key elements that allow the pathologist, examining the brain under the microscope at autopsy, to make a certain diagnosis of AD. As stated earlier, it is believed by many that the amyloid plaques are in the chain of events causal of AD.

As a result of neuronal cell death, the brain shrinks as AD progresses, and this can be seen by various types of brain scans. This is not too helpful from the diagnostic standpoint, because the brain shrinks in other kinds of dementia also. But it does give information about the progression of the disease. In general, the greater the shrinkage, the worse the dementia.

Risk Factors for Alzheimer's Disease

There are various known factors that increase the risk that an individual will get AD. One is a mutation in the amyloid precursor protein gene. This causes genetic AD, that is, a type of AD where inheritance can be shown within families. This type of AD occurs at a younger age than the much more common nongenetic form of AD, often called sporadic AD. The genetic form of AD is a relatively rare cause of AD.

A major risk factor for nongenetic AD is age. People begin getting diagnosed with AD at age 60, and the prevalence goes up with each decade of age after that. For example, while 11% of people aged 65 and older have AD, 33% of those 85 and older have the disease.[1] While more women than men have AD, this appears to be due only to the longer lifespan of women. Thus, the sex of the person is not an AD risk factor.

Family history of AD increases the risk of AD. Individuals with a first-degree relative (parent, brother, or sister) with AD have an increased risk of AD, and those with two first-degree relatives have an even higher risk.[1] The effect of family history on risk is explained by the effects of shared genes and shared environments. This effect of family history is not entirely explained by the effect of apolipoprotein E4 alleles, to be discussed next.

Certain alleles of certain special genes increase the risk of AD. These risks are detected by an increased prevalence of those alleles in AD populations versus the general population. A major example of this is the E4 allele of the apolipoprotein gene. The E2 allele of this gene seems to actually result in some protection against AD, while the E3 allele is neutral and the E4 allele increases risk. Those who have one copy of the E4 allele have a threefold greater risk of AD, while those who inherit two copies of E4 have an 8- to 12-fold greater risk.[1] It is estimated that 40%–65% of those with AD have one or two copies of an E4 allele, while about 25% of the general population have one or two copies of E4. So, overall E4 increases risk about twofold.

Two other examples of this are the "iron-management genes" hemachromatosis[4] and transferrin.[5] Mutant alleles of these two genes are found at increased frequency in AD populations, indicating they increase the risk of AD.

A more recent addition to this list of genes which influence AD risk are alleles of the ATP7B gene.[6] ATP7B is, of course, the Wilson's disease gene. Since heterozygotes for Wilson's disease causing mutations have a mildly increased body copper load, leading to increased urine and liver copper, the increased prevalence of ATP7B alleles suggest an increased body load of copper for a lifetime can be a risk factor for AD.

While these various alleles of certain genes being at increased prevalence indicate increased risk of AD, the total impact on AD prevalence is relatively small, except for apolipoprotein E4, because the frequency of the alleles in the general population is not very high.

In contrast, the impact on risk of the environmental factors described in this book, copper-2 and increased mild body copper loading in general, is huge. The prevalence of AD has increased from about 1% in the general population over age 60 in the 19th century to 10%–45% in the 20th century depending on age, largely due to these environmental factors.

THE HUMAN SIDE OF ALZHEIMER'S DISEASE

So far in this chapter the clinical picture of how AD begins, how the diagnosis is made, what is going on in the brain pathologically to explain the loss of cognition, and AD risk factors have been described. But these topics are rather dry and factual, and they do not do justice to what a human disaster AD is. A person cannot really appreciate the catastrophe the epidemic of AD is unless they understand something about the human miseries it causes. In this section, a typical example of this type of human disaster is described.

Phil W was 66, recently retired from a job at a manufacturing plant, with a decent pension including a health-care plan for his retirement. He was looking forward to an enjoyable retirement, fishing, traveling, playing golf, going to dinner with his wife, playing cards with friends, and frequent visits from his children and grandchildren. He was in reasonably good health with well-treated high blood pressure and well-treated elevation of his cholesterol. His wife Mary was also 66 and in good health. They had three adult children, James age 45, Laura age 42, and Tim age 40, all of whom had families of their own and lived a long way from their parents. They were loving children and visited their parents at least twice a year, to make sure that they and their grandchildren maintained a loving relationship with their parents and grandparents.

Mary began to notice that Phil was forgetting things. First, it was displaced objects, such as his car keys or glasses that he couldn't find. Then he started forgetting to take his pills for high blood pressure and high cholesterol. Then one day, she was shocked to learn he couldn't remember that they had dinner at a restaurant with friends the night before. Worried, she insisted he see their doctor. The doctor examined Phil, but didn't test his memory, and pronounced him healthy "for someone his age."

A few months later Mary got a call from a police officer. Phil had taken his car out to do an errand and had stopped at a local store because he couldn't remember how to get home. He also couldn't remember the home phone number. The clerk had called the police. From Phil's driver's license the officer had obtained his name, and using the phone book, called Mary. This episode led Mary to try and stop Phil from driving. Phil resisted, until one day he didn't stop at a stop sign and hit another car. Fortunately, no one was hurt, but with that Mary was able to stop his driving.

Gradually Phil could be trusted to do less and less. He could no longer play golf with his buddies because he kept wandering off. He couldn't be trusted to go fishing. He also couldn't play cards with their circle of friends because he couldn't remember how to play the various games.

Mary took Phil to their doctor, who promptly referred him to a neurologist. The neurologist did an SMMSE on him and Phil scored 14. The neurologist made the diagnosis of AD. He told Mary there was no good treatment for the disease, only a drug that would help a little temporarily. He recommended keeping Phil's mind as active as possible and lots of exercise.

About a year after Phil's problems started, son James and his family paid them a visit and Phil couldn't remember James' name, or the names of his grandchildren or daughter-in-law. About 6 months later, daughter Laura and son Tim and their families paid a joint visit and Phil didn't recognize his own children.

Meanwhile, Mary was having increasing problems taking care of Phil. He would get out of bed at night and try to walk out of the house. If he escaped, he couldn't find his way back. She had deadbolts installed on all the doors and kept the keys away from him. Nevertheless, his frequent wandering around the house at night made it difficult for her to sleep. She was constantly afraid he would get into trouble. She couldn't leave him alone in the house during the day, because he frequently tried to cook, and there was great risk of fire. So she had to take him with her for grocery shopping and other trips. But on those trips he was difficult to manage, often trying to wander off.

Meanwhile he was requiring increasing care at home, from bathing, other personal hygiene, dressing, etc. He could do none of those things on his own. Later it became necessary to feed him. He would sit in front of his food and ignore it, seeming to have forgotten that he was supposed to be eating.

Over a period of 2 years Mary became very worn down by the round-the-clock caregiving, without let up, without breaks, without vacation. The children gave moral support through frequent phone calls and emails, but they didn't offer to take Phil into their homes. They all had busy lives, and they knew how disruptive his presence could be. Mary tried to hire help, but Phil was very difficult to such employees. He didn't like these strangers and gave them a very hard time. As a consequence, the employees never lasted very long.

Then Phil became unreliable in his urinary and fecal elimination. He started urinating in the sinks, and once in the corner of the garage. Fecal passage would increasingly occur without going to the bathroom, requiring an extensive cleanup by Mary.

Three years after this all started, all three of their children came for a visit. They were shocked at how frail and tired Mary was. The family decided it was time to put Phil in a nursing home.

The next shock was how much that cost—$9000/month for the kind of round-the-clock care Phil needed. Their health insurance plan didn't include nursing home care. Mary's monthly income was $5000 from social security and Phil's pension. Her monthly expenses were about $4000, including a mortgage and car loan payments. They had no savings at this point, so Mary was forced to take out a second mortgage. Phil was in the nursing home 4 years before he died.

Think of this case from the human and emotional standpoints. For over 5 years the children and grandchildren lost their father and grandfather. He didn't know them during those 5 years, and there was no memory of the good times they had had together. Mary had the terrible burden of a 24/7 caregiver for 3 years, then a heavy financial burden for 4 years. But, worse than that, she had lost her husband for most of the 7 years, she had lost love, companionship, fun—all things spouses provide. But the worst thing, he didn't just die so he could be mourned, and then she could go on with life. He hung around 7 years, eventually not remembering her at all, for the most part not remembering anything about their lives together, and a daily reminder of all she had lost.

What about Phil? It can be argued that the AD patient loses so much cognition that they are unaware of their plight, and thus escape much suffering. It is hard to know whether this is true or not. But surely, those in the early stages of a disease, particularly if it is a common one such as AD, are aware of what the disease eventually does, through all kinds of sources of information about the disease. They must be aware that they will be reduced to complete dependence on a caregiver, like a pet, except unlike a pet, they give no joy to the caregiver. And that there will be no joy or happiness, not even awareness in their life—just day to day survival.

So Phil worked all his adult life, looking towards the joy of retirement, and then lost all his retirement years to AD. That is what is happening to a very large proportion of our elderly. This is clearly a catastrophe. AD is worse than cancer. While cancer often kills in horrible ways, at least the patient retains their ability to think, to love and be loved, to have some happiness, generally until close to the end. Not so with AD. A veil is slowly drawn over past memories and the ability to think. The person is slowly drawn into a cocoon of unawareness, eventually untouched by others and by life. While possibly they are not suffering, they are not really living either, at least not as a human. More like a pet, but not living as good a life as most pets, either. This is surely a disastrous end of life scenario for the elderly.

REFERENCES

1. Alzheimer's Disease Association. *Alzheimer's disease facts and figures*. 2016.
2. Molloy W, Caldwell P. *Alzheimer's disease*. Toronto: Key Porter Books; 1998. p. 76–8.
3. Hardy J, Selkoe DJ. The amyloid hypothesis of Alzheimer's disease. Progress and problems on the road to therapeutics. *Science* 2002;**297**:353–6.
4. Moalem S, Percy ME, Andrews DF, et al. Are hereditary hemochromatosis mutations involved in Alzheimer disease? *Am J Med Genet* 2000;**93**:58–66.
5. Zambenedetti P, DeBellis G, Biunno I, Musicco M, Zatta P. Transferrin C2 variant does confer a risk for Alzheimer's disease in Caucasians. *J Alzheimers Dis* 2003;**5**:423–7.
6. Squitti R, Polimanti R, Siotto M, et al. ATP7B variants as modulators of copper dyshomeostasis in Alzheimer's disease. *Neuromol Med* 2013;**15**:515–22.

CHAPTER THREE

Interesting and Important Historical and Demographic Facts About the Epidemic of Alzheimer's Disease in Developed Countries Pointing to Environmental Intoxicants Causing the Epidemic

These important historical and demographic facts will be listed right up front, and then the details supporting assertions that these are valid and important facts will be discussed:

1. Developing countries are in the midst of an epidemic of Alzheimer's disease (AD), with 15% of those aged 65 to 74, and 44% of those aged 75 to 84, developing AD.[1]

2. The epidemic is predominant in developed countries, with a much lower prevalence of AD in undeveloped countries. These demographics of AD prevalence are dramatically illustrated in Fig. 3.1, with a big "WHY-1?" attached, asking why this huge difference?

3. Alois Alzheimer first described the disease that now bears his name in 1907.[2] But an AD-like disease did not exist, or was very rare, before 1900. This history of the demographics is illustrated in Fig. 3.2, with the "WHY-1?" recapitulated on the right regarding current demographics, and a second "WHY-2?" at the top emphasizing the dramatic increase in prevalence in developed countries only over the last 100 years.

4. The lack of an AD-like disease of high prevalence before 1900 was not due to a lack of elderly people.

5. The lack of an AD-like disease of high prevalence before 1900 was not due to a simple attribution of the disease to normal aging, that is, a failure to notice the disease.

So according to the above, developed countries now have a major epidemic of a disease that was virtually nonexistent before 1900. These facts are

Environmental Causes and Prevention Measures for Alzheimer's Disease
ISBN 978-0-12-811162-8
http://dx.doi.org/10.1016/B978-0-12-811162-8.00003-2

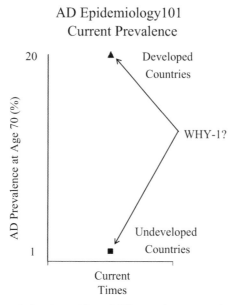

Figure 3.1 The current demographics of AD prevalence are shown, illustrating the huge difference in prevalence between developed and undeveloped countries. A major question is why the difference, hence the "WHY-1?"

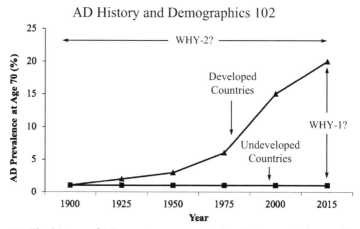

Figure 3.2 The history of AD prevalence over the last 115 years is shown, illustrating the huge increase in AD prevalence in developed, but not undeveloped, countries over that period. Added to the "WHY-1?" of current demographics is a "WHY-2?" for why the big change in prevalence, leading to the current AD epidemic in developed countries.

important (and amazing!) because they establish, practically shout out, that our epidemic of AD is due to something, or some things, newly occurring during the last century in the environment of developed countries. The facts are important because they suggest that if we identify these environmental factors, we might be able to eliminate or reduce exposure to it and thus alleviate the high prevalence of this terrible disease.

Human populations have undergone epidemics before. These are usually due to an infectious agent such as the bacteria that caused bubonic plague, or tuberculosis, or the viruses that caused smallpox or AIDS. But epidemics have also been caused by new things in the environment besides infectious agents, such as lung cancer from cigarette smoking.

Equally amazing, and surprising, as the facts listed, the Alzheimer's scientific community has not realized the importance of these facts as indicating a specific environmental causation for AD. This is probably due to, at least in part, that most in the scientific community are not even aware of the facts listed, particularly facts 2–5, and certainly people in the lay community are not aware of them. Now that a major environmental causative agent (inorganic copper as revealed in Chapter 4 and discussed further in later chapters) has been identified, it becomes important for everyone to understand these facts as background for understanding the causative agent. So each of the five assertions will be examined to see if supporting data are available to establish the assertions as valid facts.

FACT 1: ALZHEIMER'S DISEASE IS IN EPIDEMIC MODE IN DEVELOPED COUNTRIES

There is no doubt that developed countries have a huge caseload of AD currently and that the prevalence (number of cases per unit of population) has been increasing rapidly over the last 50 years. Currently in the United States, as already pointed out, about 15% of those aged 65 to 74 and 44% of those aged 75 to 84 have AD.[1] Increasingly, those who reach retirement age, have increasing memory loss, increasing inability to function, and increasing dependence on a caregiver, in other words AD, to look forward to. Fig. 3.3 provides some facts about AD in the United States, illustrating the tremendous impact of the AD epidemic.[1]

As will be shown in discussing Fact 3, AD was almost nonexistent before 1900. In 1970 there were less than 400 cases described in the world's medical literature.[3] Of course there were more cases than this in existence in

Alzheimer's Disease: Facts and Impact

- Only cause of Death in Top 10 – That Can't be Prevented, Cured or Slowed
- 6th Leading Cause of Death
- 2015 Cost - $226 Billion in U.S.
- 1 of 3 Seniors – Dies With It
- 2/3 of Cases are Women

Figure 3.3 Some facts about AD in the United States, and some of the impact of the disease, are given.[1]

1970, but this illustrates that the disease was still relatively uncommon then. Now, there are about 5 million cases of AD in the United States,[1] with an equal number with mild cognitive impairment (MCI), the precursor to AD. This explosion of cases in the last 45 years certainly qualifies the term "epidemic" for the current status of AD.

FACT 2: PREDOMINANCE OF THE AD EPIDEMIC IN DEVELOPED COUNTRIES AND NOT IN UNDEVELOPED COUNTRIES

The prevalence of AD in numerous countries has been published by Ferri et al.[4] It is clear that the current prevalences are much higher in developed countries, such as the United States, Canada, and Europe, than in undeveloped parts of the world such as Africa, South America, and India. For example, in rural India the prevalence in those aged 65 and over is 1.07%.[5] In Nigeria, Africa, the prevalence in those aged 65 to 74 is 0.5% while in those aged 75 to 84 it is 1.69%.[6] Interestingly this same study included African Americans of the same ethnic heritage as the Nigerians living in the United States, and their prevalence of AD in those aged 75 to 84 was 8.02%. This great increase in prevalence from living in the United States in the same genetic group illustrates the powerful effect of Westernization on AD prevalence.

Japan is an interesting exception. It is a developed country with a relatively low prevalence of AD.[7] Yet when Japanese migrate to Hawaii, their prevalence of AD increases to the equivalent of other developed countries.[8] This, perhaps, gives a clue about the environmental agent or agents. It or they do not appear to be in Japan, but are in Hawaii. Since a case is going to be made that an important environmental agent is exposure to inorganic copper, it will be important to see if exposure to inorganic copper is different in Japan than in Hawaii. More about that in Chapter 8, when a web of evidence incriminating inorganic copper is developed.

But for now, putting the first two facts together—there is a growing epidemic of AD, but the epidemic is primarily occurring in developed and not undeveloped countries. This should suggest that something about the environment of developed countries is causing the disease. Ordinarily when one thinks of epidemics occurring in certain areas and not in others, one thinks about infectious disease caused by a bacteria or virus that transmitted from person to person, after gaining a foothold in an area, can cause a great increase in disease frequency. But AD is not an infectious disease, so it is necessary to start thinking about other things that have changed in the environment as development occurs.

Of course many things are different in the environment of developed countries compared to undeveloped countries. One of these areas of difference is the diet. In fact, Waldman and Lamb,[3] when they wrote their book Dying for a Hamburger, postulated that beef eating, causing a prion disease like mad cow disease, was the cause of AD. Of course, beef eating occurs much more in developed countries because of the expense of buying and eating beef, but there is no supporting "web of evidence" to support their theory of a prion disease. However, it does turn out that increased meat eating, with its increased absorption of copper, is another causal environmental factor. More about that in Chapter 9.

The main point to make here is that the first two facts point to an environment agent or agents that occur in developed countries, except for Japan. Later, we will draw the noose very tightly around ingestion of inorganic copper being an important environmental agent.

FACT 3: AN ALZHEIMER'S-LIKE DISEASE WAS ALMOST ENTIRELY NONEXISTENT, OR WAS VERY RARE, BEFORE 1900

How is this known? The credit goes to Waldman and Lamb,[3] in the book already mentioned, Dying for a Hamburger. In the first part of their book, they do a marvelous job laying out the evidence for this Fact 3, and here their lead in reciting that evidence is followed.

Alois Alzheimer first described the disease that bears his name in 1907.[2] So obviously there could be no Alzheimer's disease before that, because the disease had not yet been named. But the point to be made here is that a disease like AD, if it existed at all, was very rare before 1900. For clarity, the disease before 1900 will be referred to as an AD-like disease which became AD once it was named in the early 1900s.

The first piece of evidence comes from the writings of Osler.[9] Osler, an internist physician, gathered the existing medical knowledge in the mid-1800s into a series of books. One whole volume was dedicated to diseases of the brain. There was no mention of an AD-like disease.

Second, Gowers,[10] a neurologist, wrote a textbook of neurological disease during this same period, and did not mention an AD-like disease. Third, Freud,[11] a psychiatrist, wrote extensively during this period on diseases of the brain and did not discuss an AD-like disease.

Fourth, and most compelling, Boyd,[12] a pathologist, wrote a textbook of pathology during the end of the 1800s, revised repeatedly during the early 1900s, and did not describe the amyloid plaques and neurofibrillary tangles, hallmarks of the AD brain, in the brains autopsied during that period.

Waldman and Lamb[3] cite other evidence that an AD-like disease did not exist or was rare before 1900. They review the world's literature and find no evidence of an AD-like disease before 1900. They review medical journals including Index Medicus. There was no citation on the subject of dementia until 1921. They track the number of cases recorded in the world's literature over time. From a single case in 1908, there was a slow progression upward with a rapid increase in the 1940s (150 cases). There was another large increase in recorded cases to 648 in 1978.

Thus, in summary, AD was rare or nonexistent before 1900, and began to increase slowly in the early 1900s and more rapidly in the late 1900s, until now we have about 5 million cases in the United States and an equal number of MCI cases, many converting to AD every year.

FACT 4: THE LACK OF AN AD-LIKE DISEASE OF HIGH PREVALENCE BEFORE 1900 WAS NOT DUE TO A LACK OF ELDERLY PEOPLE

A common objection to the claim that an AD-like disease was rare before 1900 is that since AD is a disease of aging, and since populations of people are steadily increasing their life spans, there were not enough elderly people back then for AD to occur and be noticed.

However, Waldman and Lamb[3] showed an age pyramid that in 1911, half the French population was living to age 60, the age at which AD currently occurs at about a 10% prevalence. To discuss the French data in a little more detail, Waldman and Lamb[3] showed an age pyramid for the 1911 population of France, on p. 71 of their book. The pyramid shows there were about 255,000 people aged 65 to 75, which at the prevalence rate of 15% published in the current Alzheimer's Disease Fact and Figures[1] would generate 52,500 patients. The pyramid shows about 115,000 people aged 75 to 85, which at the published prevalence rate of 44% would generate 50,000 cases. Thus, at current AD prevalence rates, there would have been well over 100,000 AD patients in France in 1911, which would have provided plenty of patients to be seen in clinics and at the autopsy table. The census for the US population in 1900 shows there were 3.2 million people over age 60, which would have resulted in 320,000 cases at today's rate. So if the disease prevalence was anything like today's rate in developed countries, there would have been plenty of cases to be seen by Osler,[9] Gowers,[10] and Freud[11] and to show plaques and tangles in the brains of autopsy material by Boyd.[12] So the lack of an AD-like disease before 1900 was not due to a lack of elderly people.

FACT 5: THE LACK OF AN AD-LIKE DISEASE OF HIGH PREVALENCE BEFORE 1900 WAS NOT DUE TO A SIMPLE ATTRIBUTION OF THE DISEASE TO NORMAL AGING, THAT IS, A FAILURE TO NOTICE THE DISEASE

A second common objection to the claim that an AD-like disease was rare before 1900 is that it was attributed to normal aging and that people who suffered cognition loss were thought of as just becoming senile more rapidly and did not have a special disease. Stated another way, people, including physicians, just did not notice the disease, although it was there.

It seems possible that the clinicians, Osler,[9] Gowers,[10] and Freud,[11] could have made this mistake, although it seems unlikely given the thoroughness of each one of these in their observations and descriptions. But this would not explain the failure of pathologists such as Boyd[12] to see the plaques and tangles in the brains at autopsy during the late 1800s and early 1900s.

So, summarizing, there is definitely an epidemic of AD, affecting a high proportion of people over age 60. This epidemic is affecting primarily developed countries, with undeveloped countries having a low prevalence. This epidemic is new, in the sense that it has blossomed in the last half of the last century in developed countries. The disease was essentially nonexistent, or at least very rare, before 1900. The lack of a high prevalence of this disease in the past was not due to a lack of elderly people and not due to a simple failure to recognize the disease.

If these facts are correct, and the assertion is made here that they are, it leads to the inescapable conclusion that environmental change in developed countries in the last century is causing the epidemic. It is shocking that the Alzheimer scientific community is not focusing on this and scrambling to uncover the environmental culprit. Rather, great attention and expense is being paid to develop a "disease modifying agent." Efforts are being made to develop antibodies to beta amyloid, the cause of the amyloid plaques in the brain of AD patients. But all this effort is misdirected and much less important if the causes of most cases of the disease are identified and eliminated. That is what now seems possible. A major cause of the disease is ingestion of inorganic copper, with a secondary cause being increased absorption of dietary copper from increased meat eating. Ingestion of the poisonous inorganic copper can easily be eliminated and AD prevalence greatly reduced. See Chapter 11 to learn more about this simple approach to avoiding many cases of Alzheimer's.

REFERENCES

1. Alzheimer's Disease Association. *Alzheimer's disease facts and figures*. 2016.
2. Alzheimer A. Uber eine eigenartige Erkankung der Hirnrinde. *Allg Z Psychiatr* 1907;**64**:146–8.
3. Waldman M, Lamb M. *Dying for a hamburger: modern meat processing and the epidemic of Alzheimer's disease*. New York: Thomas Dune Books/St Martin's Press; 2005.
4. Ferri CP, Prince M, Brayne C, Brodaty H, Fratiglioni L, Ganguli M, et al. Global prevalence of dementia: a Delphi Consensus Study. *Lancet* 2005;**366**:2112–7.
5. Chandra V, Ganguli M, Panda V, Johnston J, Belle S, Dekosky ST. Prevalence of Alzheimer's disease and other dementias in rural India. *Neurology* 1998;**51**:1000–8.
6. Hendrie H, Osuntokun O, Ks H, Ogunniyi AO, Hui SL, Unverzagt FW, et al. Prevalence of Alzheimer's disease and dementia in two communities: Nigerian Africans and African Americans. *Am J Psychiatry* 1995;**152**:1485–92.
7. Ueda K, Kawano H, Hasuo Y, Fujishima M. Prevalence and etiology of dementia in a Japanese community. *Stroke* 1992;**23**:798–803.
8. White L, Petrovitch H, Ross GW, Masaki KH, Abbott RD, Teng EL, et al. Prevalence of dementia in older Japanese-American men in Hawaii: the Honolulu-Asia Aging Study. *J Am Med Assoc* 1996;**276**:955–60.
9. Osler W. *Modern medicine: its theory and practice*. Philadelphia, New York: Lea and Febiger; 1910.
10. Gowers WR. *A manual of diseases of the nervous system*. Philadelphia: P. Blakiston, Son, and Co.; 1888.
11. Strachey J, Freud A, Strachey A, Tyson A. *24 volumes entitled, the standard edition of the complete psychological works of Sigmund Freud, written between 1895 and 1939*. London: The Hogarth Press and the Institute of Psycho-Analysis; 1966.
12. Boyd WA. *Textbook of pathology: an introduction to medicine*. Philadelphia: Lea and Febiger; 1938.

CHAPTER FOUR

Candidate Environmental Factors for the Alzheimer's Epidemic Part 1: The Metals—Aluminum, Lead, Mercury, Zinc, Iron, and Copper

INTRODUCTION

Oral ingestion of various metals has had a long history of being considered as causative of AD. Each one of the various suspect metals will be reviewed here, one by one, and the evidence considered as to whether a given metal could be a causative factor in the epidemic of AD.

To be identified here as highly likely to be a causative factor in the AD epidemic, the agent (here a specific metal), must pass two tests. The first is that exposure to the metal must have substantially increased in developed, but not undeveloped, countries over the last 100 years. Second, there must be substantial evidence tying the metal to the pathogenesis of AD.

The first three metals to be considered, aluminum, lead, and mercury, have three things in common that are relevant to the discussion. First, none of them have a known function in the biology of either lower organisms or in humans. Thus, when they are taken in by humans, they are simply environmental contaminants. Second, all three are neurointoxicants, that is they are neurologically damaging. Third, because they have all found uses in industrialized society, humans in developed countries have greatly increased exposure to all three over the last 100 years. Thus, all three will pass the first of our two tests, that is, a substantial increase in exposure over the last 100 years in developed countries. That makes the second test the critical one. All three of these metals can cause brain damage, even adversely affecting cognition. But the key will be do they produce the same disease picture as AD, including the brain pathology of amyloid plaques and neurofibrillary tangles?

The last three metals, zinc, iron, and copper, all share the property of being essential to life. Thus, their possible role in AD causation will be because possibly excessive amounts, or inappropriate valence species, are ingested.

Environmental Causes and Prevention Measures for Alzheimer's Disease
ISBN 978-0-12-811162-8
http://dx.doi.org/10.1016/B978-0-12-811162-8.00004-4

ALUMINUM

Aluminum (Al) is a known neurointoxicant in both animals and humans and has been suggested as causing AD for a very long time. Flaten[1] has written a very good review of the evidence. In 1965,[2] Al phosphate was injected into the brains of rabbits and produced neurofibrillary tangles reminiscent of the neurofibrillary tangles in AD, although they are not morphologically identical to those in AD. In 1973, it was reported that Al concentration was increased in the brains of AD patients.[3] In 1976,[4] Al toxicity was suggested as the cause of an encephalopathy syndrome developing in patients on dialysis for chronic renal failure.

Aluminum as the cause of the encephalopathy syndrome from dialysis is now generally accepted and is due first to a direct exposure of high concentrations of Al to the blood stream, bypassing the low absorption of Al from the intestines, and second, poor renal excretion of Al due to the kidney failure.[1] However, the brain pathology in dialysis encephalopathy is not that seen in the AD brain. Indeed, the difference in brain pathology in the two conditions has been used as an argument against Al causation of AD.[1]

There have been many studies showing an epidemiological association between higher levels of Al in drinking water and higher incidence of AD. Not all studies showed this association. Of the 13 studies reviewed by Flaten,[1] nine were positive, but four were negative for the association. In most of the positive studies, the odds ratio for risk of AD was 1.5–2.0 for high-Al levels versus lower Al levels. An example is the study of Neri and Hewitt[5] who found an odds ratio of 1.46 for water containing Al at greater than 0.2 mg/L versus water containing Al at less than 0.01 mg/L. A strength of the hypothesis of Al causation of AD is the relative consistency of positive studies of a drinking water association (9 out of 13 positive studies). A weakness is the low amount of Al even in drinking water with "high"-Al content of 0.1–0.4 mg/L in the comparisons. In contrast, the dietary intake of Al is several milligrams per day.[1] Ingestion of Al-containing antacids usually results in an intake of about 1 g of Al per day.[1] Multiple studies of antacid use (13 reviewed by Flaten[1]) have not shown a consistent pattern of increased risk of AD.[1] The negative evidence from antacid studies provides fairly convincing evidence that Al ingestion is not a risk factor for AD.[1]

To resolve the conflict between the relatively positive drinking water data and the strongly negative antacid data, it has been suggested that perhaps drinking water Al is more bioavailable or perhaps can sometimes come as a particularly neurointoxicating compound.[1] Interestingly, in the only drinking water study that examined the speciation of Al,[6] it was found that the only Al fraction that was positively associated with AD was monomeric organic Al (odds ratio of 2.67, control Al 0.012 mg/L vs. high Al of 0.077 mg/L). It is possible some organic compound of Al found in drinking water is particularly bioavailable and neurotoxic. For example, Al maltolate is an organic Al that is very neurotoxic,[7] but it has not been studied to see if it is present in drinking water.

Occupational exposure to Al usually results in inhalation of large amounts of Al. Two epidemiological studies of such workers have failed to show a consistent positive association with AD.[1]

Aluminum exposure also comes from antiperspirant use, from beverages in Al cans, and from cooking with Al pots and pans. The amount of Al taken in through these routes would be relatively small compared, for example, to antacid use, and there has been no evidence of these uses as being risk factors for AD.[1]

Examining and summarizing the evidence of a possible AD causative role of Al, the possible increase in Al in AD brains is somewhat controversial (some studies have found it, others have not). The neurotoxicity of Al produces neurofibrillary tangles, but they are not morphologically the same as the tangles in AD, and Al does not produce amyloid plaques. The brain toxicity of Al in renal dialysis encephalopathy produces a different clinical syndrome than that of AD, and it never produces amyloid plaques or neurofibrillary tangles. High doses of Al from antacid use rather clearly are not associated with a high risk for AD.

Summarizing the above data, it makes a good case that Al intake is not a risk factor for AD. However, there is the relatively good case (9 of 13 studies) of an association of drinking water Al with AD risk. Remarkably, the concentrations of Al in the "high-Al" water is about 0.2 mg/L, compared to the nontoxic effects of 1.0 g/day of Al when taken as an antacid. Thus, a dose 5000 times as high, taken as an antacid, is not a risk factor for AD, as the dose taken in drinking water, which putatively is a risk factor for AD. This comparison greatly weakens the case for the Al of drinking water as a risk factor for AD. The only remaining possibility is that a particular Al compound often found in drinking water is both very bioavailable and a very strong neurointoxicant.

Now to the two tests listed earlier for an agent to pass to be considered as highly likely to be a causative factor in the AD epidemic. Certainly Al passes the first test, in that exposure to Al has greatly increased over the last century in developed countries. However, Al essentially fails the second test. None of the data as reviewed above point to Al as important in the pathogenesis of AD. The most significant data points to a possible role of Al in drinking water in AD risk, but these data are weakened by the tiny amount of copper in drinking water versus the large amount in antacids that is not a risk factor. Further, Al in drinking water fails the first test, because it is unlikely Al levels in drinking water have increased in developed countries over the last 100 years, since the Al in drinking water seems to be derived from source water.

Another piece of negative evidence for Al as AD-causative is the animal model study of Sparks et al.[8] as discussed in the first chapter. Sparks and Schreurs[9] developed an AD animal model and found that tiny amounts of copper (0.12 ppm), added to drinking water greatly enhanced brain AD pathology and decreased the animals' memories. Sparks et al.[8] found that neither Al nor zinc added to the drinking water caused any effect of these types at all.

It is concluded that Al is ruled out as a contributor to the AD epidemic. At most, it might be a minor risk factor in drinking water.

LEAD

Lead, like Al, is a neurointoxicant. It is well known that childhood exposure leads to reduced cognitive performance and behavioral disturbances. Adult exposure leads to reduced cognition. However, in none of these human studies has it been shown that lead toxicity produces the brain pathology, amyloid plaques, and neurofibrillary tangles, of AD. A good review of the evidence supporting lead exposure as causing AD has been written by Bakulski et al.[10]

There have been positive studies in primates.[11] Monkeys exposed to lead during the first 400 days of life had adverse brain changes at 23 years of life when compared to control animals of the same age. These changes include amyloid plaques, increased amyloidogenesis, and upregulation of proteins in the amyloid producing pathway, such as amyloid precursor protein. Primates are the only known animal species to develop amyloid plaques with aging. The development of increased amyloid plaques after early life lead exposure is fairly strong evidence that early life lead exposure might be a risk factor for AD. Of course, this study has never been replicated—admittedly it is difficult to replicate a 23 year primate study. But, it is worrisome that contamination of the lead used in the studies with trace amounts of copper could explain the results. Remember from Chapter 1 that trace amounts of copper in the drinking water of animal models of AD greatly enhance brain pathology.[9] The question is, why the primates in this study show amyloid plaques in the brain from lead exposure when human exposure to lead has not shown amyloid plaques in the brain. The same group who reported the primate study reported that rats exposed to increased lead from their mother's milk during the first 20 days of life showed increased beta amyloid and increased amyloid precursor protein and its mRNA in the animals' old age.[12] These studies have not been replicated either.

Exposure to lead is relatively ubiquitous in developed countries.[10] It can come from children exposed to peeling lead-based paint in older homes. Such paint is no longer legal. Lead pipes and solder containing lead were used predominately in water systems for a long time, and some lead is leached into the drinking water. Ash from incinerator plants which deposits on the surrounding terrain may contain high concentrations of lead. Abandoned lead battery recycling plants, prevalent around the United States, and lead mines and smelters, contaminate the soil around the sites with lead. For a long time, automobile gasoline contained lead which was discharged into the air and settled onto the land.

Many of these sources of lead exposure have been strongly curtailed in recent times in developed countries. In the United States for example, mean blood level in 1978 was 12.8 μg/dL but was reduced to 2.8 by 1991.[10] Since the "action" blood level of lead for lead poisoning in children is set at 10 μg/dL, it can be seen that the mean level of 12.8 μg/dL in 1978 meant that much of the population was exposed to excess lead. Interestingly though, the AD prevalence has not plummeted since 1991, which might be expected if lead was playing a strongly causative role. However, it may be too early to see an effect.

Regarding the two tests for lead causation of the AD epidemic, lead passes the first test, because exposure to lead has certainly increased greatly over the last 100 years in developed countries, and probably not so much in undeveloped countries. However, lead does not pass the second test. Except for the studies of one group in primates,[11] there is no evidence that lead toxicity produces the brain pathology, amyloid plaques, and neurofibrillary tangles, of AD. The work of the group reporting lead causation of amyloid brain pathology in primates has not been replicated, and it is possible the pathology in primates is due to copper contamination. Importantly, while lead exposure in humans has been very prevalent, amyloid plaque pathology has not been reported. Further, there appears to be no data correlating increased AD prevalence with increased lead exposure in subgroups known to have been heavily exposed.

It is concluded that lead is not a factor in the causation of the AD epidemic. However, based on the animal studies which show increased amyloid precursor protein, beta amyloid, and even increased amyloid plaques, in the brain from lead exposure,[11,12] lead exposure, particularly early in life, may be a risk factor for AD, although probably a minor one. Based on the animal studies, lead exposure early in development may be particularly important.

MERCURY

Mercury is extremely toxic, and the toxicology is well known.[13] Mutter et al.[14] have written a good review of the possible causative role of mercury in AD.

There are various forms of mercury. Metallic mercury (Hg^0) is the only metal that is liquid at room temperature, and it evaporates easily.[14] Because it is uncharged it is very diffusible and lipid soluble. It penetrates cells easily and also the blood–brain barrier. Intracellularly, the relatively inactive Hg^0 is quickly oxidized to Hg^{++}, which is a strong oxidant, producing oxidative stress and cell damage. The brain is the major target organ for Hg^0.[14]

Another form of mercury is organic mercury. Methylmercury is one common form. Extracellular thiol groups facilitate the absorption of organic mercury. Intracellularly, organic mercury is converted to inorganic mercury that is Hg^+ or Hg^{++}, which again causes oxidant damage.

The potential risk of mercury for AD comes from first, gaseous Hg^0 in the brain, and second, the conversion of methylmercury to Hg^{++} in the brain. A major source of Hg^0 is from dental amalgam fillings. The retention of Hg^0 from this source ranges from 1 to 22 µg/day, the actual number depending on several factors, including the number and recency of fillings.[14] A major source of organic mercury is fish eating.[15] If a person ingests one meal of fish per week, the average retention of methylmercury is about 2.3 µg.

Mercury exposure can also come from working in certain kinds of jobs and professions. Types of factories that expose workers to mercury include chlorine–alkaline battery and thermometer factories, mercury extraction operations, and occasionally gold mining.[14] And, of course, dentists and dental workers are also exposed because they are installing amalgam fillings.

Studies in mercury exposed workers clearly demonstrate that mercury exposure is an occupational health hazard. There are definite negative effects on memory, attention, and cognition. These are relevant to AD, but there are other symptoms such as sleep disturbance, mood swings, and pain which are not relevant. There is a correlation between blood and other levels of mercury with the severity of these symptoms. While cognition is affected, it is not clear it progresses to dementia from mercury exposure. Studies attempting to relate health effects from dental amalgams have been largely negative.[14]

Studies of mercury content in AD brains at autopsy have given mixed results, some showing mercury accumulation and others not.[14] It seems clear that mercury accumulation in the AD brain is not a consistent finding. Animal studies of mercury exposure produced some of the biochemical evidence of damage seen in AD brains but never showed amyloid plaques or neurofibrillary tangles.

It is concluded that while mercury exposure passes the first of our two tests, namely that exposure has greatly increased in the last 100 years in developed countries, it does not pass the second test. There is no substantial evidence that mercury is playing a significant role in AD pathogenesis in most AD patients. At most it might be contributing to cognitive decline in patients developing AD from other causes. It is concluded that mercury is not an environmental factor contributing to the AD epidemic.

ZINC

Zinc is an extremely important element in the body, binding to many proteins and conferring catalytic and/or structural properties. It is the most abundant trace metal in the brain, where it plays many roles.[16]

Regarding zinc's possible role in AD, Watt et al.[16] have published an informative review. First of all, zinc is found in amyloid plaques.[17] Second, it has been shown that Aβ (beta amyloid) can be caused to aggregate by zinc,[18] but the concentration of zinc required varies widely among various studies,[19] and it appears somewhat unlikely that the physiological concentration of available (chelatable, or free) zinc is high enough for this to occur.

The concentration of zinc in the AD brain is diminished compared to age-matched controls, according to most studies, particularly in some regions of the brain.[20] Burnet[21] hypothesized that zinc deficiency was a possible cause of dementia, and Constantinidis[22] proposed a specific zinc deficiency in the central nervous system of AD patients could be partially causative of the disease. More recently it has been clearly shown that AD patients are systemically zinc deficient based on serum levels.[23] A possible mechanism is poor nutrition.

Zinc has several protective roles in the brain. An important one is, it is secreted with glutamate into the synapse of glutaminergic neurons and shuts down excitation, thus protecting against excitotoxicity.[20] Based on the hypothesis of zinc deficiency in AD, Constantinidis[24] tried zinc therapy in an uncontrolled study in AD patients and reported excellent efficacy. Later, zinc supplementation was given in an AD animal model and caused significant improvement in the animals.[25] Most recently, it was shown that zinc therapy in a small double-blind trial in AD patients, halted cognition loss significantly in those aged 70 and over compared to placebo controls.[26]

In summary, there was a period when the finding of zinc in amyloid plaques and the demonstration in vitro that zinc could cause Aβ aggregation, that zinc was considered as possibly partly causative of AD. However, in more recent times, the finding of zinc deficiency in AD and the likely beneficial effect of zinc therapy has caused thinking to shift away from zinc toxicity playing a causal role.

An important study that supports this latter conclusion was carried out by Adlard et al.[27] In this study they knocked out the zinc transporter-3 (ZnT-3) gene of mice. ZnT3 is a zinc transporter in the brain responsible for loading zinc into the vesicle in the neuron that releases zinc into the synapse. The knockout of ZnT3 caused cognitive loss in the mice and was called "a phenocopy for the synaptic and memory deficits of Alzheimer's disease." In this publication, they also added some insight as to what might be causing neuronal zinc deficiency. They suggest the amyloid plaques, by binding zinc, deprive the rest of the brain from having adequate zinc. This genetic production of neuronal zinc deficiency recapitulating AD adds the clinching piece of data that zinc deficiency, not zinc toxicity, is potentially causal in AD. However, if another piece of data is needed, in the study of Sparks et al.[8] in which tiny amounts of copper added to the drinking water of animal models of AD greatly accelerated the disease, zinc (and aluminum), did not.

Turning to the two tests for causality, there is reason to believe that zinc exposure may be higher in the last 100 years, because of the use of zinc supplement pills and increased meat eating, since zinc is better absorbed from meat than from vegetable foods. Thinking in the reverse, there is no reason to suspect that zinc deficiency has increased in the last 100 years in developed countries. So zinc fails the first test, in the sense there is better evidence for a causal role for zinc deficiency than zinc toxicity, and no increase in zinc deficiency in the last 100 years. Zinc toxicity also fails the second test, since it has become apparent that there is no evidence that zinc plays a causal role in the pathogenesis of AD. Zinc deficiency may play some role, but it would appear to be minor. In conclusion, it is apparent that zinc is not causal of the AD epidemic.

IRON

Iron has been a suspect as causative of AD for some time. It has been shown that iron accumulates in the amyloid plaques of AD brains.[17] Iron, like copper, is a transition element, meaning it is redox active, transitioning between the oxidized, Fe^{+++} state, and the reduced, Fe^{++} state. The body takes advantage of this to catalyze many reactions important to life. However, a side effect of redox reactions is the generation of toxic reactive oxygen species (ROS), which can cause oxidant damage.

Iron is prevalent in the human brain and is very important in brain metabolism. But because of its redox activity and its accumulation in amyloid plaques, it has long been suspected as causing neurodegeneration in AD.[28,29] It is known that when iron[30,31] as well as copper[31] bind to amyloid plaques, they generate damaging ROS.[32] Oxidant damage is considered to be an important contributing factor to neurodegeneration in AD.[30,32]

A more recent study with advanced technology has localized the buildup of iron and related it to tissue damage in the AD brain.[33] The hippocampus of the brain is an important site for developing memory and other cognitive functions and is known to be involved early in AD, while the thalamus of the brain is not involved early. In the study they showed that the hippocampus accumulated iron early in AD and showed tissue damage in correlation with iron accumulation, while the thalamus did not accumulate iron early and did not show tissue damage.[33]

Summarizing, there seems little doubt that iron contributes to the oxidant damage in AD once amyloid plaques are formed. But the evidence is that iron contributes to ROS generation by binding to preformed plaques.[30,31] If one accepts the amyloid cascade hypothesis for AD causation, something has to trigger plaque formation in the beginning, and there is no evidence iron can do that.

Regarding the two tests for a role for iron in the causation of the AD epidemic in developed countries, with respect to the first, there is good evidence that people in developed countries have considerably more exposure to iron in the last 100 years. This is because iron is much better absorbed from meat than from vegetable foods, and it is clear that in developed countries, meat eating has greatly increased in the last 100 years. This has not happened in undeveloped countries for economic reasons.

However, iron fails the second test, of a pathogenic path to cause AD. According to the most widely accepted theory of AD causation, the aggregation of beta amyloid into amyloid plaques starts the disease. Iron does not appear to do this. Iron could be a factor in aggravating the disease, once initiated, by producing ROS and oxidant damage. There is a type of contrary evidence to the conclusion that iron is not causative of AD. Mutant alleles of the hemochromatosis gene[34] and the transferrin gene,[35] both genes which can influence iron levels, are more common in AD populations than in the general population. This indicates that possessing these alleles increases the risk of AD, which in turn suggests that increased iron levels can be causative of AD. However, these alleles are relatively rare in the general population, so they could not be responsible for the great increase in AD prevalence seen in developed countries in the last 100 years.

That iron is contributing to the toxicity in the AD brain is supported by a therapeutic trial of the iron chelator, desferrioxamine, given for 2 years to AD patients.[36] The trial was very positive in terms of slowing the clinical progression of dementia.

Thus, in conclusion, while there is little question that iron contributes to the severity of AD, and may have a minor effect on causation, it is also clear that it is not a major factor in causing the AD epidemic in developed countries.

COPPER

Copper has been suspected for some time as at least partially causative of AD based first, on its capability to cause aggregation of beta amyloid into amyloid plaques,[37] and second, on causing release of damaging oxidant radicals when binding to amyloid plaques.[31] As has been mentioned before, oxidant damage is believed to be a key pathogenic element in AD. Like iron, copper is a transition element, meaning it can transition from the oxidized, Cu^{++} state, to the reduced, Cu^+ state. Again like iron, this has been taken advantage of in biological metabolism to catalyze important reactions, but it also makes it vulnerable to produce ROS and oxidant damage.

A major event propelled copper forward as a causative agent for AD. That was the study and paper by Sparks and Schruers,[9] discussed in detail in Chapter 1, in which they showed that tiny amounts of copper, 0.12 ppm, added to the drinking water of a rabbit model of AD greatly enhanced amyloid plaque formation in the rabbit brain and enhanced memory loss in the animals. This study is unique in that it shows that copper actually enhances amyloid plaque formation. No other putative causative agent has been shown to do that. In fact, Sparks et al.[8] went on to show that two other candidate causative agents, aluminum and zinc, did not cause increased plaque formation, or memory loss, in the same models that showed AD-type damage from copper. They also showed that copper caused the same worsening of AD-like disease in several other AD animal models, besides the rabbit model.[8] Their findings regarding copper were later confirmed in another laboratory.[38] An increase of this tiny amount of copper in the animal food would have no toxicity. In fact, copper in animal chow can be increased from 3 to 6 ppm, a 25-fold greater increase than 0.12 ppm, and there is no toxicity.

The 2003 study of Sparks and Schreurs[9] was followed in 2006 by a very important publication by Morris et al.[39] This group studied cognition in a large group of people over time in relation to intake of various nutrients and supplements. They found that those in the highest quintile of copper intake, who were there because they took a copper supplement pill, if they also ate a high-fat diet, as many people in developed countries do, lost cognition at six times the rate of other groups. This is a dramatic observation. Taking copper in a supplement pill, and probably half of Americans take a multimineral/multivitamin pill all of which contain copper, damages cognition at an alarming rate.

Brewer, realizing that the drinking water copper in the Sparks and Schreurs[9] study and the copper in supplement pills in the Morris et al.[39] study were both inorganic copper as opposed to the organic copper in food, developed the hypothesis that inorganic copper was causative, or at least a trigger of AD.[26,40-48] Plausibility of this hypothesis is increased by the observation made many years ago that some of the inorganic copper labeled with copper-64 appeared in the blood in 1–2 h.[49] If the organic copper in food is radioactively labeled, the labeled copper does not appear in the blood for days, and then as part of proteins secreted by the liver into the blood. Obviously some inorganic copper bypasses the liver and appears immediately in the blood, where it is hypothesized to be toxic to cognition.

Further support for the concept of special toxicity of inorganic copper came from studies by Ceko et al.[50] This group studied the speciation of copper in food and water, meaning they studied the relative amounts of Cu^{++} and Cu^+ in these substances. They found, as expected, that copper in water was Cu^{++}, but surprisingly, they found that copper in various foods was primarily Cu^+. This latter result was unexpected because in life, Cu^{++} and Cu^+ form a redox doublet, critical for many reactions important to metabolism. But apparently at death, or harvest, in the absence of oxygen transport, Cu^{++} is reduced to Cu^+. This means that humans and their ancestors evolved primarily ingesting Cu^+ and not Cu^{++}. This explains the presence in the gastrointestinal track of a system for handling Cu^+ but not Cu^{++}. The intestinal transporter Ctrl[51] will transport Cu^+ but not Cu^{++}. The Cu^+ is transported via a system that transports it to the liver, which puts the copper into safe channels. Cu^{++} cannot be absorbed through Ctrl. It can be absorbed through other systems or by diffusion and some of it bypasses the liver and is absorbed directly into the blood, where it may be toxic to cognition. The inorganic copper hypothesis is also known as the copper-2 hypothesis, since copper-2 is another designation for Cu^{++}.

Further impetus to the toxicity of copper in AD is provided by the work of the Squitti group. These workers have shown that the pathogenesis of AD is closely tied to the size of the blood free copper pool. The blood copper can be thought of as in two pools: in one, the copper is covalently bound to ceruloplasmin. In the second pool, copper is loosely bound to albumin and some other molecules. This second pool of copper is called free copper, although it is not really free, just more loosely bound. It is this copper that is readily available for use by organs of the body and that is potentially toxic if it becomes expanded, as it is in a disease called Wilson's disease.[52] Wilson's disease is an inherited disease of copper accumulation and copper toxicity, in which the blood free copper pool is greatly increased. In AD, the Squitti group has shown that the free copper pool is significantly increased,[53] is correlated to cognition measures,[54] to cognition loss over time,[55] and predicts the conversion of MCI patients, the precursor state to AD, to full AD.[56] This work ties the pathogenesis of AD closely to the size of the blood free copper pool and indicts copper toxicity as very important in AD.

In more recent work, the Squitti group has shown another copper-related result. They find that carriers of certain mutant alleles of the ATP7B gene are increased in prevalence in the AD population.[57] The ATP7B gene is the Wilson's disease gene, which as just stated, is an inherited disease of copper toxicity. Wilson's disease is an autosomal recessive disease, meaning both copies of the gene have to be mutated to produce the disease. Carriers of one Wilson's disease allele have mild elevations of body copper, not requiring treatment. The finding that carriers of ATP7B alleles are increased in prevalence in the AD population indicates that these alleles increase the risk of developing AD. If it is assumed that these alleles increase the body load of copper a little, it suggests that mildly increasing the body copper load increases the risk of AD. The paradox here is why homozygous Wilson's disease, with a huge increase in body copper load, causes a neurologic movement disorder but no effect on cognition, while AD does not cause a neurologic movement disorder, but damages cognition. The answer may be that homozygotes with Wilson's disease die young or are treated young, eliminating the excess copper load, while carriers live a lifetime with a mildly increased copper load, allowing age, a major risk factor for AD, to interact with increased copper to produce the disease.

If increased prevalence of ATP7B alleles in the AD population means an increased copper load is a risk factor for AD, the question comes up, is there any reason to suspect increased general copper loading, irrespective of copper valence, in the populations of developed countries during the last 100 years? The answer is yes. As already discussed in the iron section of this chapter, there has been a great increase in meat eating in developed countries in the last century due to the increased raising of animals for meat, and copper, as well as iron, is much better absorbed from meat than from vegetable foods.[58] It is estimated that absorption of copper is around 50% greater from meat than from vegetable sources. Undeveloped countries have not shared in the great increase in meat eating for economic reasons.

If inorganic copper, or copper-2, is evaluated as a separate AD risk factor, independent of total body copper load, is there evidence that ingestion of copper-2 has greatly increased in developed countries over the last 100 years? The answer again is yes. The ingestion of copper supplement pills, in which all of the copper is copper-2, has greatly increased in developed but not undeveloped countries. Perhaps more important is the increased use of copper plumbing in developed countries over the last century. The epidemic of AD, developing over the last 100 years and exploding in the last 50 years, parallels closely the "epidemic" of copper plumbing use during those periods.[59] (More on that in Chapters 6 and 8.) It has been shown that enough copper leaches from copper plumbing into the drinking water in one-third to two-thirds of North American homes to cause AD,[26] if the animal model studies of Sparks and Schreurs[9] are a good guide.

Turning to our two tests for the risk factor as causing the AD epidemic of the last century, in the last two paragraphs we have spelled out evidence that both copper-2 exposure and general copper absorption have both greatly increased over the last century in developed countries, but not in undeveloped countries. So both copper-2 and general copper overload from meat eating pass the first test.

Both also pass the second test. The animal model studies show that tiny amounts of copper-2 (inorganic copper) cause greatly enhanced amyloid plaque development and memory loss.[9] Much greater ingestion of copper-1 (organic copper) would not do this. Copper-2 is the only metal, or other AD putative causative agent, to cause amyloid plaques to form. Regarding general copper overload, the increased prevalence of ATP7B alleles in AD populations indicates that copper load is an AD risk factor, and increased meat eating provides a mechanism for general increase in copper load.

CONCLUSIONS REGARDING METALS AS CAUSATIVE OF AD EPIDEMIC

It is concluded that Al, lead, mercury, iron, and zinc are excluded from being important causes of the AD epidemic in developed countries over the last 100 years. While some of them, particularly iron, may contribute to worsening of AD once initiated, none of these five metals pass the two tests of being major AD causative factors.

The situation with copper is quite different. It is concluded that copper-2 ingestion is an excellent candidate as a major causative factor in the AD epidemic in developed countries. Second, it is concluded that an increase in copper absorption in general from increased meat eating, irrespective of copper valence, is also an excellent candidate as a major causative factor in the AD epidemic in developed countries.

REFERENCES

1. Flaten TP. Aluminium as a risk factor in Alzheimer's disease, with emphasis on drinking water. *Brain Res Bull* 2001;**55**:187–96.
2. Klatzo I, Wisniewski H, Streicher E. Experimental production of neurofibrillary degeneration. I. Light microscopic observations. *J Neuropathol Exp Neurol* 1965;**24**:187–99.
3. Crapper DR, Krishnan SS, Dalson AJ. Brain aluminum distribution in Alzheimer's disease and experimental neurofibrillary degeneration. *Science* 1973;**180**:511–3.
4. Alfrey AC, LeGendre GR, Kaehny WD. The dialysis encephalopathy syndrome: possible aluminum intoxication. *N Engl J Med* 1976;**194**:184–8.
5. Neri LC, Hewitt D. Aluminium, Alzheimer's disease, and drinking water. *Lancet* 1991;**338**:390.
6. Gauthier E, Fortier I, Courchesne F, Pepin P, Mortimer J, Gauvreau D. Aluminum forms in drinking water and risk of Alzheimer's disease. *Environ Res* 2000;**84**:234–46.
7. Nelson WO, Lutz TG, Orvig C. The chemistry of neurologically active, neutral, and water soluble aluminum complexes. In: Lewis TE, editor. *Environmental chemistry and toxicology of aluminum*. Chelsea (MI): Lewis Publishers; 1989. p. 271–87.
8. Sparks DL, Friedland R, Petanceska S, Schreurs BG, et al. Trace copper levels in drinking water, but not zinc or aluminum, influence CNS Alzheimer-like pathology. *J Nutr Health Aging* 2006;**10**:247–54.

9. Sparks DL, Schreurs BG. Trace amounts of copper in water induce beta-amyloid plaques and learning deficits in a rabbit model of Alzheimer's disease. *Proc Natl Acad Sci USA* 2003;**100**:11065–9.

10. Bakulski KM, Rozek LS, Dolinoy DC, Paulson HL, Howard H. Alzheimer's disease and environmental exposure to lead: the epidemiologic evidence and potential role of epigenetics. *Curr Alzheimer Res* 2012;**9**:563–73.

11. Wu J, Basha MR, Brock B, et al. Alzheimer's disease (AD)-like pathology in aged monkeys after infantile exposure to environmental metal lead (Pb): evidence for a developmental origin and environmental link for AD. *J Neurosci* 2008;**18**:3–9.

12. Basha MR, Wei W, Bakheet SA, et al. The fetal basis of amyloidogenesis: exposure to lead and latent overexpression of amyloid precursor protein and beta-amyloid in the aging brain. *J Neurosci* 2005;**25**:823–9.

13. Clarkson TW, Magos L, Myers GJ. The toxicology of mercury – current exposures and clinical manifestations. *N Engl J Med* 2003;**349**:1731–7.

14. Mutter J, Naumann J, Sadaghiani C, Schneider R, Wallach H. Alzheimer disease: mercury as pathogenetic factor and apolipoprotein E as a modulator. *Neuroendocrinol Lett* 2004;**25**:275–83.

15. World Health Organization. *Health risks of heavy metals from long-range transboundary airpollution.* Copenhagen: WHO Regional Office for Europe; 2007.

16. Watt NT, Whitehouse IJ, Hooper NM. The role of zinc in Alzheimer's disease. *Int J Alzheimers Dis* 2011:971021.

17. Lovell MA, Robertson JD, Teesdale WJ, Campbell JL, Markesbery WR. Copper, iron and zinc in Alzheimer's disease senile plaques. *J Neurol Sci* 1998;**158**:47–52.

18. Bush AI, Pettingell WH, Multhaup G, et al. Rapid induction of Alzheimer Aβ amyloid formation by zinc. *Science* 1994;**265**:1464–7.

19. Faller P, Hureau C. Bioinorganic chemistry of copper and zinc ions coordinated to amyloid- β peptide. *Dalton Trans* 2009;**7**:1080–94.

20. Cuajungco MP, Lees GJ. Prevention of zinc neurotoxicity in vivo by N,N,N′,N′–tetrakis(2-pyridylmethyl)ethylenediamine (TPEN). *Neuroreport* 1996;**7**:1301–4.

21. Burnet FM. A possible role of zinc in the pathology of dementia. *Lancet* 1981;**1**:186–8.

22. Constantinidis J. Alzheimer's disease: the zinc theory. *Encephale* 1990;**16**:231–9.

23. Ventriglia M, Brewer GJ, Simmonelli I, Mariani S, Siotto M, Bucossi S, et al. Zinc in Alzheimer's disease: a meta-analysis of serum, plasma, and cerebrospinal fluid studies. *J Alzheimers Dis* 2015;**46**:75–87.

24. Constantinidis J. Treatment of Alzheimer's disease by zinc compounds. *Drug Dev Res* 1992;**27**:1–14.

25. Corona C, Masciopents F, Silvestri E, et al. Dietary zinc supplementation of 3xTg-AD mice increases BDNF levels and prevents cognitive deficits as well as mitochondrial dysfunction. *Cell Death Dis* 2010;**1**:e91.

26. Brewer GJ. Copper excess, zinc deficiency, and cognition loss in Alzheimer's disease. *Biofactors* 2012;**38**:107–13.

27. Adlard PA, Parncutt JM, Finkelstein DI, Bush AI, Cash AD. Cognitive loss in zinc transporter-3 knock-out mice: a phenocopy for the synaptic and memory deficits of Alzheimer's disease? *J Neurosci* 2010;**30**:1631–6.

28. Perry G, Sayre LM, Atwood CS, Castellani RJ. The role of iron and copper in the aetiology of neurodegenerative disorders: therapeutic implications. *CNS Drugs* 2002;**16**:339–52.

29. Bishop GM, Robinson S, Liu Q, Perry G. Iron: a pathological mediator of Alzheimer disease? *Dev Neurosci* 2002;**24**:184–7.

30. Smith MA, Harris PLR, Sayre LM, et al. Iron accumulation in Alzheimer disease is a source of redox generated free radicals. *Proc Natl Acad Sci USA* 1997;**94**:9866–8.

31. Sayre LM, Perry G, Harris PL, Liu Y, et al. In situ oxidative catalysis by neurofibrillary tangles and senile plaques in Alzheimer'1s disease: a central role for bound transition metals. *J Neurochem* 2000;**74**:270–9.
32. Casadesus G, Smith MA, Zhu X, et al. Alzheimer disease: evidence for a central pathogenic role of iron mediated reactive oxygen species. *J Alzheimer Dis* 2004;**6**:165–9.
33. Raven EP, Lu PH, Tishler TA, Heydari P, Bartozokis G. Increased iron levels and decreased tissue integrity in hippocampus of Alzheimer's disease detected in vivo with magnetic resonance imaging. *J Alzheimer Dis* 2013;**37**:127–36.
34. Moalem S, Percy ME, Andrews DF, et al. Are hereditary hemochomatosis mutations involved in Alzheimer disease? *Am J Med Genet* 2000;**93**:58–66.
35. Namekata K, Imagawa M, Terashi A, Ohta S, Oyama F, Ilhara Y. Association of transferrin C2 allele with late-onset Alzheimer's disease. *Hum Genet* 1997;**101**:126–9.
36. Crapper McLachlan DR, Dalton AJ, Kruck TP, et al. Intramuscular desferrioxamine in patients with Alzheimer's disease. *Lancet* 1991;**337**:1304–8.
37. Sarell CJ, Wilkinson S, Viles JH. Sub-stoichiometric levels of copper ions accelerate the kinetics of fibre formation and promote cell toxicity of amyloid beta from Alzheimer's disease. *J Biol Chem* 2010;**285**:41533–40.
38. Singh I, Sagare AP, Coma M, Perlmutter D. Low levels of copper disrupt brain amyloid-beta homeostasis by altering its production and clearance. *Proc Natl Acad Sci USA* 2013;**110**:14471–6.
39. Morris MC, Evans DA, Tangney CC, Bienias JL. Dietary copper and high saturated and trans fat intakes associated with cognitive decline. *Arch Neurol* 2006;**63**:1085–8.
40. Brewer GJ. Elevated levels of dietary copper may accelerate cognitive decline and hasten the onset of Alzheimer disease. *Nutr MD* 2007;**33**:1–4.
41. Brewer GJ. The risks of free copper in the body and the development of useful anticopper drugs. *Curr Opin Clin Nutr Metab Care* 2008;**11**:727–32.
42. Brewer GJ. The risks of copper toxicity contributing to cognitive decline in the aging population and to Alzheimer's disease. *J Am Coll Nutr* 2009;**28**:238–42.
43. Brewer GJ. The risks of copper and iron toxicity during aging in humans. *Chem Res Toxicol* 2010;**23**:319–26.
44. Brewer GJ. Issues raised involving the copper hypotheses in the causation of Alzheimer's disease. *Int J Alzheimers Dis* 2011;**99**:1–11.
45. Brewer GJ. Copper toxicity in Alzheimer's disease. Cognitive loss from ingestion of inorganic copper. *J Trace Elem Med Biol* 2012;**26**:89–92.
46. Brewer GJ. Alzheimer's disease causation by copper toxicity and treatment with zinc. *Front Aging Neurosci* 2014;**6**:92.
47. Brewer GJ, Kaur S. Ingestion of inorganic copper from drinking water and supplements is a major factor in the epidemic of Alzheimer's disease. In: Rothkopf MM, editor. *Metabolic medicine and surgery.* Boca Raton (FL): CRC Press; 2014.
48. Brewer GJ. Divalent copper as a major triggering agent in Alzheimer's disease. *J Alzheimers Dis* 2015;**46**:593–604.
49. Hill GM, Brewer GJ, Juni JE, Prasad AS. Treatment of Wilson's disease with zinc. II. Validation of oral 64 copper with copper balance. *Am J Med Sci* 1986;**292**:344–9.
50. Ceko MJ, Aitken JB, Harris HH. Speciation of copper in range of food types by x-ray absorption spectroscopy. *Food Chem* 2014;**164**:50–4.
51. Ohink H, Thiele DJ. How copper traverses cellular membranes through the copper transporter 1, Ctrl. *Ann NY Acad Sci* 2013;**1314**:32–41.
52. Brewer GJ. Wilson's disease. In: Kasper DL, As F, Hauser SL, Longo DL, Jameson JL, Loscalzo J, editors. *Harrison's principles of internal medicine.* 19th ed. New York: McGraw-Hill Companies, Inc.; 2015.
53. Squitti R, Pasqualetti P, Dal Forno G, Moffa F, et al. Excess of serum copper not related to ceruloplasmin in Alzheimer disease. *Neurology* 2006;**64**:1040–6.

54. Squitti R, Barbati G, Rossi L, Ventriglia M. Excess of nonceruloplasmin serum copper in AD correlates with MMSE, CSF [beta]-amyloid, and h-tau. *Neurology* 2006;**67**:76–82.
55. Squitti R, Bressi F, Pasqualetti P, Bonomini C. Longitudinal prognostic value of serum "free" copper in patients with Alzheimer disease. *Neurology* 2009;**72**:50–5.
56. Squitti R, Ghidoni R, Siotto M, Ventriglia M. Value of serum nonceruloplasmin copper for prediction of mild cognitive impairment conversion to Alzheimer disease. *Ann Neurol* 2014;**75**:574–80.
57. Squitti R, Polimanti R, Siotto M, Bucossi S. ATP7B variants as modulators of copper dyshomeostasis in Alzheimer's disease. *Neuromol Med* 2013;**15**:515–22.
58. Brewer GJ, Yuzbasiyan-Gurkan V, Dick R, Wang Y. Does a vegetarian diet control Wilson's disease? *J Am Coll Nutr* 1993;**12**:527–30.
59. Foley PT. *International copper demand patterns – the case of plumbing tube*. New York (NY): CRU Consultants Inc.; 1985. p. 183–6. Economics of Internationally Traded Minerals, Economics of Copper, Section 5.2.

CHAPTER FIVE

Candidate Environmental Factors for the Alzheimer's Epidemic Part 2: Diet and Other Lifestyle Factors

INTRODUCTION

In this chapter, the search will be continued for environmental risk factors that could explain, or at least help to explain, the epidemic of Alzheimer's disease (AD) that has occurred in developed, but not undeveloped, countries in the last century. In the last one or two decades much has been written about how diet, various nutritional factors, exercise, and even behavioral modifications like yoga and meditation can mitigate the severity, or even reduce the risk, of AD. While these factors may be useful and even important in the management of AD, in the context of this chapter it must be remembered that the search is for factors that can help explain the AD epidemic. To meet this role it must be postulated, for example in the case of a nutrient that is beneficial to AD, that the nutrient was available until about 100 years ago, and then became much less available in developed, but not undeveloped, countries during the last 100 years, thus causing a major increase in AD prevalence in developed countries in the last 100 years. Keeping in mind the search is not for factors that cause a minor increase in AD prevalence but for those that cause a major increase in prevalence, it appears it will be rather difficult for most factors considered here to fulfill the necessary criteria. In this chapter we will consider diet in general, specific nutrients, exercise, and behavioral modifications. As in Chapter 4, to be considered as a significant causative factor of the AD epidemic the factor under consideration must pass the two tests. First its presence must be strongly different in developed countries, in the 19th, compared to the 20th, centuries. Second, there must be strong evidence that ties the factor strongly to the pathogenesis of AD.

GENERAL DIET

Increased Meat and Fat Ingestion

In Chapter 4, meat ingestion came up in the context of both copper and iron being much better absorbed from meat than from vegetable foods.

Environmental Causes and Prevention Measures for Alzheimer's Disease
ISBN 978-0-12-811162-8
http://dx.doi.org/10.1016/B978-0-12-811162-8.00005-6

Since both copper and iron are transition elements that can generate reactive oxygen species (ROS) and oxidant damage, and oxidant damage is an underlying event in AD, increased meat eating could be an AD causative factor. Further supporting the concept of increased iron and copper as causative of AD, both the iron management genes hemochromatosis[1] and transferrin[2] and the copper control gene ATP7B[3] have variant alleles of increased frequency in AD populations, indicating increased iron and copper levels both increase AD risk.

If meat ingestion is increased, it follows that fat ingestion will also be automatically increased. A high fat intake seems to be intimately tied to copper toxicity. In the original study of Sparks and Schreurs,[4] the AD model was a high-cholesterol-fed animal model in which tiny amounts of copper in the drinking water greatly enhanced AD. (In later studies,[5] tiny amounts of copper greatly enhanced AD without special feeding.) In the study of Morris et al.[6] people in the highest quintile of copper intake suffered greatly enhanced cognition loss, if they also ingested a high-fat diet. Grant[7] has published a paper in which it is found that average fat intake of the population of a country correlate positively with AD prevalence in the country.

Regarding the two tests for a factor to be considered as possibly causative of the AD epidemic, increased meat and fat eating definitely pass the first test. Developed countries in the last century have greatly increased raising animals for their meat, resulting in a great increase in average meat and fat ingestion.

Regarding the second test, increased meat eating can be tied to the pathogenesis of AD through increased copper and iron absorption from meat eating.[8] It has been estimated that 50% more copper is absorbed in a diet with meat as opposed to a vegetarian diet. In Chapter 4, copper was identified as a likely causative factor in the AD epidemic, so greatly increased meat eating, through copper increase, also becomes a likely causative factor. Fat and copper work together synergistically to damage neurons. Copper can oxidize some fats and cholesterol into molecules especially damaging to neurons.

In summary, increased meat and fat ingestion in developed countries in the last century are accepted as likely significant contributors to the AD epidemic.

Increased Refined Carbohydrate Ingestion

This topic deals with the great increase in ingestion of refined carbohydrates that are quickly metabolized, such as the sugars sucrose, fructose, and glucose. There is no question that intake of refined carbohydrates has greatly increased in developed countries in the last century. The intake of soft drinks sweetened with fructose, the intake of sucrose as a sweetener, and the intake of refined sugars in general have greatly increased in developed, but not so much in undeveloped countries. The high intake of these sugars has been blamed for causing, at least in part, the "metabolic syndrome" epidemic in developed countries in the last century. This syndrome includes the epidemics of obesity, type-2 diabetes, and hypertension that have plagued developed countries.

However, the only known connection to AD is that diabetes is a mild risk factor for AD. This risk factor does not seem to be strong enough to be much of a causal factor in the AD epidemic, so in summary processed carbohydrate ingestion passes the first test, but not the second test, as a likely causative factor for the AD epidemic.

Decreased Fruit and Vegetable Ingestion

It follows that if meat and refined carbohydrate ingestion are increased in developed countries in the last century, other parts of the diet such as fruits and vegetable ingestion, have decreased. Since the intake of more fruits and vegetables is generally considered as healthful, has the relative decrease in these important parts of the diet been a significant factor in the AD epidemic in developed countries in the last century?

In the next section we will consider specific potentially protective substances in the diet many of which come from fruits and vegetables, and whether their absence, or relative decrease, could be a factor in the AD epidemic in developed countries. Here, we are considering fruits and vegetables more broadly.

The question of the possible effect of reduced intake of fruits and vegetables as partially causative of AD is made somewhat complicated by the positive effect of meat eating on AD prevalence. Since increased intake of meat, shown to be a significant causative factor of AD earlier in this chapter, is also associated with a decreased intake of fruits and vegetables, there will be a positive correlation between AD prevalence and decreased fruits and vegetable intake due to the positive effect of meat eating on AD prevalence.

As will be shown in the next section, there are specific substances in fruits and vegetables that can mitigate AD. But so far, there has been no evidence that a general relative reduction in fruits and vegetables is causative of AD. For example, there is no evidence that being a vegetarian in a developed country protects against AD.

THE EFFECT OF SPECIFIC NUTRIENTS AND FOODS

Over the last many years, specific nutrients and foods have been suggested as having potentially beneficial effects in AD. These substances are being considered here because, at least theoretically, if they have a beneficial effect, their absence might have a causative effect. So, to pass the test of causality and be considered as partially causative of the AD epidemic in developed countries, there must be reasonable evidence that the nutrient or food is strongly beneficial to AD patients, good evidence that the nutrient was present in the diet in substantial amounts up until 1900, and good evidence it was lacking in the diet in developed, but not undeveloped countries since 1900. And meeting these criteria only passes the first of the two tests set out in Chapter 4, that of relative excess or scarcity during the right time frame in developed countries. The substance must also pass the second test, that its absence can be tied closely to AD pathogenesis.

Table 5.1 provides a list of nutrients and foods that have been put forward with some evidence that the substance is beneficial in AD. Also listed in the table are the types of evidence in support of benefit in AD, along with references. The types of evidence include first, the kind of observations that often trigger interest in the substance, either high intake is related to decreased risk or low intake is related to increased risk of AD. Six of the 10 substances of Table 5.1 had evidence of this type. Positive AD animal model studies were used as supportive in 9 of the 10 substances presented in Table 5.1. In vitro studies, by which is usually meant various types of cell culture studies, were done and were supportive in 5 of the 10 cases. In one case (vitamin D) genetic studies showed positive results. Finally, a type of clinical trial was done with three of the substances (Table 5.1). None of these clinical trials were randomized placebo controlled trials (RCTs) which provide a very high standard of evidence. All the trials reported in Table 5.1 are uncontrolled studies where the substance is given, and benefit is reported. While results of these can be used to suggest more definitive studies should be done, they are very far from providing proof the substance is efficacious. They are very subject to what is called "the placebo effect" where benefit may be due to what the patient believes or hopes is happening.

These nutrients and foods will each be briefly discussed, along with an analysis of whether ingestion of the substance substantially changed in developed countries in the last 100 years. A good review of the potential benefits in AD of various nutrients and foods is given in Rege et al.[9]

Table 5.1 Data Supporting Specific Nutrients or Foods Reducing the Risk of AD or Mitigating the Disease

Nutrient or Food	Intake Related to Risk of AD	Positive Animal Studies	Positive In Vitro Studies	Positive Clinical Trials	Positive Genetic Data	References
Vitamin D	X	X		X	X	10–18
Vitamin E	X	X				19–21
Vitamin C	X					9,22
B Vitamins (B6,B12, Folate)	X	X		X		23–28
Omega Fatty Acids	X	X	X			29–32
Epigallocatechin (ECCG)		X	X			33–36
Curcumin		X	X	X		37–39
Resveratrol		X	X			40–44
Walnuts		X	X			45–47
Coffee	X	X				48–50

Vitamins

Vitamin D is obtained through the diet or is synthesized in the skin during exposure to ultraviolet rays from sunlight. The active form of vitamin D, 1-25-hydroxyvitamin D3, binds to the vitamin D receptor (VDR) and this is important to cognition.[10] Genetic variation in the VDR gene can reduce the affinity of VDR for vitamin D_3, which leads to neuronal damage and increases risk for AD.[11,12] An association has been found between low levels of plasma vitamin D and poor cognitive function[13] and cognitive decline,[14] while higher levels of vitamin D intake is associated with a better cognitive test performance in AD patients[15] and with a lower risk of AD.[16] Supplementation studies with vitamin D have shown an improvement in cognitive function.[17] Efficacy was shown by vitamin D supplementation in a mouse model of AD.[18]

Summarizing, from the above it is clear that vitamin D is important to neuronal health, and its deficiency leads to ill effects, including cognition loss and increased risk of AD. So it is clear that vitamin D passes the second of the two tests for AD epidemic causation, that is, its deficiency is intimately tied to cognition loss. To pass the first test, there has to be a relatively severe deficiency of vitamin D in developed countries only during the last 100 years relative to previous times. As mentioned, vitamin D comes from the diet and from exposure to sunlight. There is no reason to suspect a major change in sunlight exposure during the last 100 years in developed countries. While it is true there have been increased inside jobs, such as in factories and desk jobs, with a decrease in agricultural work with its increased sun exposure, this is probably offset by increased outside leisure activities, such as swimming, sunbathing, and sports. And the diet would not be expected to produce less vitamin D absorption in developed countries in the last 100 years. In addition, increased supplementation with vitamin D would offset any small decrease. In summary, it is concluded that vitamin D deficiency is not causative of the AD epidemic in developed countries.

Vitamin E is an antioxidant, and since it is clear that in AD the brain suffers from oxidative damage, it is only natural that it would be investigated as playing a protective role against AD. A population study in Chicago showed an association between higher dietary intake of vitamin E and reduced incidence of AD and cognitive decline.[19] Another study in the Netherlands showed the same thing.[20] An animal model study showed that reduction of vitamin E levels caused increased amyloid plaque formation in the brain.[21]

Vitamin C is also a strong antioxidant. A population study showed increased vitamin C levels was associated with a decreased risk of AD.[22] However, uncontrolled studies of vitamin C supplementation in AD have yielded inconsistent results.[9]

The B vitamins, here including vitamin B6, vitamin B12, and folic acid, will be considered together. Low serum concentrations of vitamin B12 or folic acid are associated with increased risk of AD[23] in one study. In another, higher intakes of folic acid and vitamin B6 were associated with a lower risk of AD.[24] In another, high folic acid intake showed reduced risk of AD,[25] and in yet another, high vitamin B12 reduced AD risk.[26] AD animal model studies of all three B vitamins have been positive.[27,28]

Summarizing the situation for vitamin E, vitamin C, and the B vitamins, all are clearly beneficial, and low levels or deficiency increase the risk for AD. Thus, they all pass the second test. However, there is no evidence that a relative deficiency of any of them in the last 100 years compared to previous times has occurred. Therefore, it is concluded that none are candidates to be causative of the AD epidemic.

Omega-3-Fatty Acids

Omega-3 fatty acids (ω-3) are essential fatty acids that are not synthesized by the human and must be obtained from the diet. The most important source is fish oil from fatty fish. About a third of the phospholipids in the gray matter of the brain is docosahexaenoic (DHA) from fish oil, so these fatty acids are very important in the brain.[9] There is a close association between DHA intake and decreased risk of AD.[29] DHA also showed efficacy in AD animal models[30,31] and showed appropriate AD-protective effects in in vitro studies.[32]

Summarizing, ω-3 are obviously important to brain health and in decreasing risk of AD, but there is nothing to suggest that there has been a substantial decrease in ω-3 ingestion in the last century. So ω-3 does not pass the first test and is not causative of the AD epidemic.

Polyphenols

The next three nutrients are polyphenols, which are antioxidants and quite available nutrients in the diet of most people. The three considered here have been shown to play roles in neuroprotection. They play protective roles in AD as antioxidants and antiinflammatory agents, as well as possibly through some other mechanisms.[9]

Epigallocatechin-3-gallate (ECCG) is a phenolic molecule with a high level in green tea. It has potent antioxidant properties.[9] Supplementation with ECCG has good efficacy in AD animal models.[33,34] In vitro studies have shown that it regulates enzymes involved in processing amyloid precursor protein[33,35] and suppresses formation of and destabilizes beta amyloid fibrils.[36]

Curcumin is a polyphenolic substance that is extracted from the Indian curry spice turmeric. It also has strong antioxidant and antiinflammatory properties.[9] In a mouse model of AD, where curcumin was supplemented for 6 months, strong reduction in inflammatory markers, and alteration of amyloid plaque formation were seen.[37] In vitro, curcumin was effective against beta amyloid aggregation.[38] And in a clinical trial, curcumin showed excellent efficacy in behavioral and cognitive symptoms in AD patients.[39]

Resveratrol is a natural polyphenol found in the skin and seeds of grapes and berries, and in numerous plant products such as peanuts, grains, and tea.[40] It shows efficacy in AD mouse models[41,42] as well as potentially beneficial effects in in vitro studies.[43,44]

Walnuts

Walnuts also have strong antioxidant and antiinflammatory properties. In a ranking of over 1000 foods according to antioxidant activity, walnuts came in second.[45] Supplementation with walnuts showed efficacy in a mouse model of AD.[46] In vitro studies of walnut extract also found positive against beta amyloid induced cellular damage.[47]

Coffee

Coffee is an excellent source of caffeine, and caffeine appears to have a variety of protective effects on the brain. A population study in women showed high coffee consumption offered protective effects against cognition decline over 4 years.[48] Caffeine was also effective in AD mouse models.[49,50]

Summary of the Role in the AD Epidemic of Specific Nutrients and Foods

All of the nutrients and foods discussed in this section share the property of showing efficacy in AD, in terms of either reducing risk of AD, or of decreasing AD severity in AD patients and/or AD animal models, and/or both reducing risk and decreasing severity. So they all potentially pass the second test of possible causation of the AD epidemic in developed countries, in that a case could be made for the protective effect of each against AD pathogenesis.

Regarding the first test, in the relevant sections, the vitamins and ω-3 fatty acids were discussed and ruled out as potentially causative of the AD epidemic because there is no evidence to support a lack of these nutrients in developed countries in the last 100 years relative to the time before that. The same argument applies to the last five substances discussed, polyphenols (ECCG, curcumin, and resveratrol), walnuts, and coffee. While, all are beneficial in some way in AD, there is no evidence that their intake has been greatly reduced in the last 100 years in developed countries compared to prior times. Thus, all the nutrients and foods considered in this section are ruled out as factors in the AD epidemic in developed countries.

PHYSICAL ACTIVITY—ANIMAL MODEL STUDIES
AD-type Brain Pathology

A number of AD animal model studies have shown the beneficial effects of exercise on amyloid pathology. In one AD mouse model with an amyloid precursor protein mutation, voluntary wheel running enhanced beta amyloid processing.[51] Studies in AD animal models have shown that exercise reduces the beta amyloid burden.[52–54]

Hyperphosphorylation of the protein, tau, is an important brain pathological hallmark of AD. Hyperphosphorylated tau causes the neurofibrillary tangles seen in the AD brain. Exercise in AD animal models has been effective in reducing tau phosphorylation.[55–58]

Behavior and Cognition

There are a large number of AD animal model studies that show that exercise improves behavior and/or cognitive abilities of the animals.[52,54,59–70]

In summary, it is clear that, based on AD animal models, exercise shows a strongly beneficial role on both AD-type brain pathology and also on the behavioral and cognitive sequelae of the disease.

PHYSICAL ACTIVITY—HUMAN STUDIES

There are many fewer studies of the effect of exercise in human AD patients compared to AD animal models. No studies were found on the effect of exercise on AD brain pathology in humans. There are a few studies that show beneficial effects of exercise on symptoms or cognition in either AD or in aged patients. One study showed that AD patients who performed regular exercise had fewer neuropsychiatric symptoms.[71] Learning and memory were enhanced in older people by increased physical activity.[72–74]

COMBINATION OF LIFESTYLE CHANGES AND AD RISK, OR TREATMENT OF AD

There are a few studies that indicate a combination of exercise and a Mediterranean or other healthy diet may decrease risk of AD.[75–77]

One anecdotal study by Bredesen employed a protocol involving 25 changes or treatments.[78]

These involved diet, reducing stress, optimizing sleep, exercise, brain stimulation, treatment with various vitamins, herbs, other agents, and hormones as necessary. There was some customizing of the protocol—patients could opt out of some elements and some were deemed unnecessary in a specific patient. Ten patients with AD, mild cognitive impairment (MCI), or what was called subjective cognitive impairment were treated. Nine of the 10 patients showed improvement in cognition within 3–6 months, the lone failure being in a patient with late and advanced AD. Six of the nine patients had lost their jobs or were struggling with their jobs, and all six regained full working status. Improvements had been sustained for up to two and a half years. These results are anecdotal, that is, the study had no controls.

SUMMARY AND CONCLUSIONS

It was concluded in the General Diet, Increase of Meat and Fat Ingestion section that increased meat and fat ingestion meets the two tests of potentially contributing to the AD epidemic in developed countries in the last 100 years. This conclusion is based on the greatly increased absorption of copper from meat as opposed to vegetable foods, and copper was found to be a causative factor in Chapter 4. The role of increased fat is that fats and copper work together to form oxidized metabolites that are toxic to neurons.

It was concluded that other than meat and fats, various general changes in the diet were not causative of the AD epidemic. Similarly, a long list of specific nutrients and foods were found not causative of the AD epidemic. While most of them had beneficial effects in AD patients, none were found to be so greatly changed in intake in the last century to be potentially causative of the AD epidemic.

Regarding physical activity, it is clear from AD animal model studies that exercise has beneficial effects on AD-type brain pathology and AD-type symptoms. The evidence is unclear in humans whether exercise effects AD risk or ameliorates symptom severity. There is one interesting but anecdotal study that suggests that a combination of lifestyle changes and various herbal and other treatments can greatly ameliorate AD.

It is concluded that exercise passes the second test of possibly being involved in pathogenesis of AD, but as with all the specific nutrients evaluated earlier, it does not pass the first test. It is unlikely that exercise and physical activity in developed countries in the last century are adequately different from previous time to account for the epidemic. While life has become more sedentary in developed countries, there is no strong evidence that sedentary people are at a far greater risk of AD than more active people.

In summary, increased meat and fat ingestion are the only lifestyle factors that are concluded to be causative of the AD epidemic in developed countries.

REFERENCES

1. Moalem S, Percy ME, Andrews DF, et al. Are hereditary hemochromatosis mutations involved in Alzheimer disease? *Am J Med Genet* 2000;**93**:58–66.
2. Zambenedetti P, De Bellis G, Biunno I, Musicco M, Zatta P. Transferrin C2 variant does confer a risk for Alzheimer's disease in Caucasians. *J Alzheimers Dis* 2003;**5**:423–7.
3. Squitti R, Polimanti R, Siotto M, et al. ATP7B variants as modulators of copper dyshomeostasis in Alzheimer's disease. *Neuromol Med* 2013;**15**:515–22.

4. Sparks DL, Schreurs BG. Trace amounts of copper in water induce beta-amyloid plaques and learning deficits in a rabbit model of Alzheimer's disease. *Proc Natl Acad Sci USA* 2003;**100**:11065–9.
5. Sparks DL, Friedland R, Petanceska S, et al. Trace copper levels in the drinking water, but not zinc or aluminum, influence CNS Alzheimer-like pathology. *J Nutr Health Aging* 2006;**10**:247–54.
6. Morris MC, Evans DA, Tangney CC, et al. Dietary copper and high saturated and trans fat intakes associated with cognitive decline. *Arch Neurol* 2006;**63**:1085–8.
7. Grant WB. Dietary links to Alzheimer's disease. *Alzheimers Dis Rev* 1997;**2**:42–55.
8. Brewer GJ, Yuzbasiyan-Gurkan V, Dick R, Wang Y. Does a vegetarian diet control Wilson's disease? *J Am Coll Nutr* 1993;**12**(5):527–30.
9. Rege SD, Geetha T, Broderick TL, Babu JR. Can diet and physical activity limit Alzheimer's disease risk?. *Curr Alzheimer Res* 2016; Mar;**14**. PMID: 26971938 [ahead of print].
10. Buell JS, Dawson-Hughes B. Vitamin D and neurocognitive dysfunction: preventing "D"ecline? *Mol Aspects Med* 2008;**29**:415–22.
11. Gezen-Ak D, Dursun E, Ertan T, et al. Association between vitamin D receptor gene polymorphism and Alzheimer's disease. *Tohoku J Exp Med* 2007;**212**:275–82.
12. Lehmann DJ, Refsum H, Warden DR, Medway C, Wilcock GK, Smith AD. The vitamin D receptor gene is associated with Alzheimer's disease. *Neurosci Lett* 2011;**504**:79–82.
13. Annweiler C, Llewellyn DJ, Beauchet O. Low serum vitamin D concentrations in Alzheimer's disease: a systematic review and meta analysis. *J Alzheimers Dis* 2013;**33**:659–74.
14. Annweiler C, Schott AM, Allali G, et al. Association of vitamin D deficiency with cognitive impairment in older women: cross-sectional study. *Neurology* 2010;**74**:27–32.
15. Oudshoorn C, Mattace-Raso FUS, van der Velde N, Colin EM, van der Cammen TJM. Higher serum vitamin D3 levels are associated with better cognitive test performance in patients with Alzheimer's disease. *Dement Geriatr Cogn Disord* 2008;**25**:539–43.
16. Annweiler C, Rolland Y, Schott AM, et al. Higher vitamin D dietary intake is associated with lower risk of Alzheimer's disease: a 7-year follow-up. *J Gerontol A Biol Sci Med Sci* 2012;**67**:1205–11.
17. Annweiler C, Beauchet O. Vitamin D-mentia: randomized clinical trials should be the next step. *Neuroepidemiology* 2011;**37**:249–58.
18. Durk MR, Han K, Chow ECY, et al. 1a,25-Dihydroxyvitami n D3 reduces cerebral amyloid-13 accumulation and improves cognition in mouse models of Alzheimer's disease. *J Neurosci* 2014;**34**:7091–101.
19. Morris MC, Evans DA, Tangney CC, et al. Relation of the tocopherol forms to incident Alzheimer disease and to cognitive change. *Am J Clin Nutr* 2005;**81**:508–14.
20. Devore EE, Grodstein F, van Rooij FJA, et al. Dietary antioxidants and long-term risk of dementia. *Arch Neurol* 2010;**67**:819–25.
21. Nishida Y, Ito S, Ohtsuki S, et al. Depletion of vitamin E increases amyloid beta accumulation by decreasing its clearances from brain and blood in a mouse model of Alzheimer disease. *J Biol Chem* 2009;**284**:33400–8.
22. Engelhart MJ, Ml G, Ruitenberg A, et al. Dietary intake of antioxidants and risk of Alzheimer disease. *JAMA* 2002;**287**:3223–9.
23. Wang HX, Wahlin A, Basun H, Fastbom J, Winblad B, Fratiglioni L. Vitamin B(12) and folate in relation to the development of Alzheimer's disease. *Neurology* 2001;**56**:1188–94.
24. Corrada MM, Kawas CH, Hallfrisch J, Muller D, Brockmeyer R. Reduced risk of Alzheimer's disease with high folate intake: the Baltimore Longitudinal Study of Aging. *Alzheimers Dement J Alzheimers Assoc* 2005;**1**:11–8.
25. Luchsinger JA, Tang M-X, Miller J, Green R, Mayeux R. Relation of higher folate intake to lower risk of Alzheimer disease in the elderly. *Arch Neurol* 2007;**64**:86–92.

26. Hooshmand B, Solomon A, Kareholt I, et al. Homocysteine and holotranscobalamin and the risk of Alzheimer disease: a longitudinal study. *Neurology* 2010;**75**:1408–14.
27. Wei W, Liu Y-H, Zhang C-E, et al. Folate/vitamin-B 12 prevents chronic hyperhomocysteinemia-induced tau hyperphosphorylation and memory deficits in aged rats. *J Alzheimers Dis* 2011;**27**:639–50.
28. Zhuo J-M, Pratico D. Acceleration of brain amyloidosis in an Alzheimer's disease mouse model by a folate, vitamin B6 and B12-deficient diet. *Exp Gerontol* 2010;**45**:195–201.
29. Grimm MOW, Kuchenbecker J, Grosgen S, et al. Docosahexaenoic acid reduces amyloid beta production via multiple pleiotropic mechanisms. *J Biol Chem* 2011;**286**:14028–39.
30. Arsenault D, Julien C, Tremblay C, Calon F. DHA improves cognition and prevents dysfunction of entorhinal cortex neurons in 3xTg-AD mice. *PLoS One* 2011;**6**:e17397.
31. Green KN, Martinez-Coria H, Khashwji H, et al. Dietary docosahexaenoic acid and docosapentaenoic acid ameliorate amyloid-beta and tau pathology via a mechanism involving presenilin1 levels. *J Neurosci* 2007;**27**:4385–95.
32. Hashimoto M, Tozawa R, Katakura M, et al. Protective effects of prescription n-3 fatty acids against impairment of spatial cognitive learning ability in amyloid J3-infused rats. *Food Funct* 2011;**2**:386–94.
33. Rezai-Zadeh K, Shytle D, Sun N, et al. Green tea epigallocatechin-3-gallate (EGCG) modulates amyloid precursor protein cleavage and reduces cerebral amyloidosis in Alzheimer transgenic mice. *J Neurosci* 2005;**25**:8807–14.
34. Rezai-Zadeh K, Arendash GW, Hou H, et al. Green tea epigallocatechin-3-gallate (EGCG) reduces beta-amyloid mediated cognitive impairment and modulates tau pathology in Alzheimer transgenic mice. *Brain Res* 2008;**1214**:177–87.
35. Ono K, Yoshiike Y, Takashima A, Hasegawa K, Naiki H, Yamada M. Potent anti-amyloidogenic and fibril-destabilizing effects of polyphenols in vitro: implications for the prevention and therapeutics of Alzheimer's disease. *J Neurochem* 2003;**87**:172–81.
36. Millington C, Sonego S, Karunaweera N, et al. Chronic neuroinflammation in Alzheimer's disease: new perspectives on animal models and promising candidate drugs. *BioMed Res Int* 2014;**2014**:309129.
37. Lim GP, Chu T, Yang F, Beech W, Frautschy SA, Cole GM. The curry spice curcumin reduces oxidative damage and amyloid pathology in an Alzheimer transgenic mouse. *J Neurosci* 2001;**21**:8370–7.
38. Cole GM, Teter B, Frautschy SA. Neuroprotective effects of curcumin. *Adv Exp Med Biol* 2007;**595**:197–212.
39. Hishikawa N, Takahashi Y, Amakusa Y, et al. Effects of turmeric on Alzheimer's disease with behavioral and psychological symptoms of dementia. *AYU* 2012;**33**:499–504.
40. Rege SD, Kumar S, Wilson ON, et al. Resveratrol protects the brain of obese mice from oxidative damage. *Oxid Med Cell Longev* 2013;**2013**:419092.
41. Kim HJ, Lee KW, Lee HJ. Protective effects of piceatannol against betaamyloid-induced neuronal cell death. *Ann NY Acad Sci* 2007;**1095**:473–82.
42. Porquet D, Casadesus G, Bayod S, et al. Dietary resveratrol prevents Alzheimer's markers and increases life span in SAMP8. *Age Dordr Neth* 2013;**35**:1851–65.
43. Feng X, Liang N, Zhu D, et al. Resveratrol inhibits 13-amyloid-induced neuronal apoptosis through regulation of S1RT1-ROCK1 signaling pathway. *PLoS One* 2013;**8**:e59888.
44. Vingtdeux V, Giliberto L, Zhao H, et al. AMP-activated protein kinase signaling activation by resveratrol modulates amyloid-beta peptide metabolism. *J Biol Chem* 2010;**285**:9100–13.
45. Halvorsen BL, Carlsen MH, Phillips KM, et al. Content of redox-active compounds (i.e., antioxidants) in foods consumed in the United States. *Am J Clin Nutr* 2006;**84**:95–135.
46. Muthaiyah B, Essa MM, Lee M, Chauhan V, Kaur K, Chauhan A. Dietary supplementation of walnuts improves memory deficits and learning skills in transgenic mouse model of Alzheimer's disease. *J Alzheimers Dis* 2014;**42**:1397–405.

47. Muthaiyah B, Essa MM, Chauhan V, Chauhan A. Protective effects of walnut extract against amyloid beta peptide-induced cell death and oxidative stress in PC12 cells. *Neurochem Res* 2011;**36**:2096–103.
48. Ritchie K, Carriere I, de Mendonca A, et al. The neuroprotective effects of caffeine: a prospective population study (the Three City Study). *Neurology* 2007;**69**:536–45.
49. Arendash GW, Schleif W, Rezai-Zadeh K, et al. Caffeine protects Alzheimer's mice against cognitive impairment and reduces brain beta-amyloid production. *Neuroscience* 2006;**142**:941–52.
50. Laurent C, Eddarkaoui S, Derisbourg M, et al. Beneficial effects of caffeine in a transgenic model of Alzheimer's disease-like tau pathology. *Neurobiol Aging* 2014;**35**(2079):90.
51. Wolf SA, Kronenberg G, Lehmann K, et al. Cognitive and physical activity differently modulate disease progression in the amyloid precursor protein (APP)-23 model of Alzheimer's disease. *Biol Psychiatry* 2006;**60**:1314–23.
52. Yuede CM, Zimmerman SD, Dong H, et al. Effects of voluntary and forced exercise on plaque deposition, hippocampal volume, and behavior in the Tg2576 mouse model of Alzheimer's disease. *Neurobiol Dis* 2009;**35**:426–32.
53. Nichol KE, Poon WW, Al P, Cribbs DH, Glabe CG, Cotman CW. Exercise alters the immune profile in Tg2576 Alzheimer mice toward a response coincident with improved cognitive performance and decreased amyloid. *J Neuroinflammation* 2008;**5**:13.
54. Adlard PA, Perreau VM, Pop V, Cotman CW. Voluntary exercise decreases amyloid load in a transgenic model of Alzheimer's disease. *J Neurosci* 2005;**25**:4217–21.
55. Stranahan AM, Martin B, Maudsley S. Anti-inflammatory effects of physical activity in relationship to improved cognitive status in humans and mouse models of Alzheimer's disease. *Curr Alzheimer Res* 2012;**9**:86–92.
56. Leem Y-H, Lim H-J, Shim S-B, Cho J-Y, Kim B-S, Han P-L. Repression of tau hyperphosphorylation by chronic endurance exercise in aged transgenic mouse model of tauopathies. *J Neurosci Res* 2009;**87**:2561–70.
57. Liu H, Zhao G, Zhang H, Shi L. Long-term treadmill exercise inhibits the progression of Alzheimer's disease-like neuropathology in the hippocampus of APP/PS1 transgenic mice. *Behav Brain Res* 2013;**256**:261–72.
58. Um H-S, Kang E-B, Koo J-H, et al. Treadmill exercise represses neuronal cell death in an aged transgenic mouse model of Alzheimer's disease. *Neurosci Res* 2011;**69**:161–73.
59. Souza LC, Filho CB, Goes ATR, et al. Neuroprotective effect of physical exercise in a mouse model of Alzheimer's disease induced by amyloid peptide. *Neurotox Res* 2013;**24**:148–63.
60. Pareja-Galeano H, Brioche T, Sanchis-Gomar F, et al. Effects of physical exercise on cognitive alterations and oxidative stress in an APP/PSN1 transgenic model of Alzheimer's disease. *Rev Esp Geriatr Gerontol* 2012;**47**:198–204.
61. Garcia-Mesa Y, Lopez-Ramos JC, Gimenez-Llort L, et al. Physical exercise protects against Alzheimer's disease in 3xTg-AD mice. *J Alzheimers Dis* 2011;**24**:421–54.
62. Belarbi K, Schindowski K, Burnout S, et al. Early Tau pathology involving the septohippocampal pathway in a Tau transgenic model: relevance to Alzheimer's disease. *Curr Alzheimer Res* 2009;**6**:152–7.
63. Hyde LA, Kazdoba TM, Grilli M, et al. Age-progressing cognitive impairments and neuropathology in transgenic CRND8 mice. *Behav Brain Res* 2005;**160**:344–55.
64. Ke H-C, Huang H-J, Liang K-C, Hsieh-Li HM. Selective improvement of cognitive function in adult and aged APP/PS1 transgenic mice by continuous non-shock treadmill exercise. *Brain Res* 2011;**1403**:1–11.
65. Van Praag H, Christie BR, Sejnowski TJ, Gage FH. Running enhances neurogenesis, learning, and long-term potentiation in mice. *Proc Natl Acad Sci USA* 1999;**96**:13427–31.
66. Van Praag H, Shubert T, Zhao C, Gage FH. Exercise enhances learning and hippocampal neurogenesis in aged mice. *J Neurosci* 2005;**25**:8680–5.

67. Dao AT, Zagaar MA, Levine AT, Salim S, Eriksen JL, Alkadhi KA. Treadmill exercise prevents learning and memory impairment in Alzheimer's disease-like pathology. *Curr Alzheimer Res* 2013;**10**:507–15.
68. Parachikova A, Nichol KE, Cotman CW. Short-term exercise in aged Tg2576 mice alters neuroinflammation and improves cognition. *Neurobiol Dis* 2008;**30**:121–9.
69. Liu H, Zhao G, Cai K, Zhao H, Shi L. Treadmill exercise prevents decline in spatial learning and memory in APP/PS1 transgenic mice through improvement of hippocampal long-term potentiation. *Behav Brain Res* 2011;**218**:308–14.
70. Hoveida R, Alaei H, Oryan S, Parivar K, Reisi P. Treadmill running improves spatial memory in an animal model of Alzheimer's disease. *Behav Brain Res* 2011;**216**:270–4.
71. Christofoletti G, Oliani MM, Bucken-Gobbi LT, Gobbi S, Beinotti F, Stella F. Physical activity attenuates neuropsychiatric disturbances and caregiver burden in patients with dementia. *Clin Sao Paulo Braz* 2011;**66**:613–8.
72. Colcombe S, Kramer AF. Fitness effects on the cognitive function of older adults: a meta-analytic study. *Psychol Sci* 2003;**14**:125–30.
73. Kang JH, Manson JE, Breteler MMB, Ware JH, Grodstein F. Physical activity, including walking, and cognitive function in older women. *JAMA* 2004;**292**:1454–61.
74. Middleton LE, Mitnitski A, Fallah N, Kirkland SA, Rockwood K. Changes in cognition and mortality in relation to exercise in late life: a population based study. *PLoS One* 2008;**3**:e3124.
75. Scarmeas N, Luchsinger JA, Schupf N, et al. Physical activity, diet, and risk of Alzheimer's disease. *JAMA* 2009;**302**:627–37.
76. Mattson MP. Pathways towards and away from: Alzheimer's disease. *Nature* 2004;**430**:631–9.
77. Mayeux R. Epidemiology of neurodegeneration. *Annu Rev Neurosci* 2003;**26**:81–104.
78. Bredesen DE. Reversal of cognitive decline: a novel therapeutic program. *Aging* 2014;**6**(9):707–17.

CHAPTER SIX

Identification of Copper-2 and Copper in General, as Major Environmental Intoxicants in the Alzheimer's Disease Epidemic: The Copper Hypothesis

In Chapters 4 and 5, candidate agents for causing the Alzheimer's disease (AD) epidemic in developed countries over the last 100 years were carefully examined. (Relevant referencing will be found in Chapters 4 and 5.) Chapter 4 dealt with candidate metals. This area was chosen for review because various metals have been put forward as AD causative over the last three or four decades. In the cases of aluminum, lead, and mercury, all three of which are pollutants and not useful in the body, the reasons for their candidacy was that all three are neurointoxicants, that is, capable of causing neurodegeneration, and exposure to all three has greatly increased in developed countries over the last 100 years. However, none of the three reproduce the AD-specific brain pathology of amyloid plaques and neurofibrillary tangles and were ruled out at being significant factors in the AD epidemic.

Iron was considered in Chapter 4, because iron level is increased in amyloid plaques, and it is redox active, capable of causing oxidant damage, and oxidant damage is a known toxicity in the AD brain. Exposure to iron has increased in developed countries over the last 100 years because of the great increase in meat eating, and iron intake and absorption is greatly enhanced from meat eating. Certain alleles of the hemochromatosis and transferrin genes, which have the potential to increase iron levels, are at increased frequency in AD populations and suggest that increased iron levels are a risk for AD. However, these alleles are too rare to be causative of the AD epidemic. And iron does not appear to cause the formation of amyloid plaques. It only binds to them once formed and then causes increased oxidant radical release. And vegetarianism in developed countries does not appear to be a major protective factor against AD. So iron has been ruled out as a major causative factor in the AD epidemic, although it may be a minor risk factor.

Environmental Causes and Prevention Measures for Alzheimer's Disease
ISBN 978-0-12-811162-8
http://dx.doi.org/10.1016/B978-0-12-811162-8.00006-8

Zinc was a candidate for being AD causative because it is increased in amyloid plaques and can cause formation of beta amyloid aggregates in vitro. But conditions in vivo are not adequate for zinc to cause plaque formation, and it is now clear that zinc deficiency is present in AD and that zinc therapy may be beneficial. So zinc has been ruled out as a causative of the AD epidemic.

Copper was found to be a major causative factor in the AD epidemic in Chapter 4. The evidence incriminating copper will be reviewed and discussed a little later in this chapter, and the following two chapters will focus on copper.

The diet and other lifestyle factors as possible causative agents in the AD epidemic were examined in Chapter 5. The reason for this examination was the multiple examples of nutrients and foods, as well as exercise, that have been shown to be mitigating factors in AD patients. The rationale that follows from this is that if an agent can reduce the severity or reduce the risk of AD, perhaps it was plentiful before 1900, but then was greatly reduced in the last 100 years, and because it was very important in preventing AD, its absence was AD causative.

In Chapter 5 it was concluded that increased meat eating in developed countries was AD causative because it greatly increased the absorption of copper, already shown in Chapter 4, to be AD causative. Data from Italy have shown that possession of an ATP7B allele, which increases body copper loading, increases risk for AD. It was concluded in Chapter 5 that increased meat eating is a contributing factor to the AD epidemic in developed countries, because of increased copper absorption and increased body copper loading.

Other than meat eating, other dietary changes and specific nutrients and foods failed to meet both of the tests for being causative of the AD epidemic. While all of the specific nutrients and foods examined could cause improvement in the severity of AD, a case could not be made for any of them that they were more plentifully ingested before 1900 than in the last 100 years in developed countries. The same was true of physical exercise. While increased exercise could show some amelioration of AD, there is no reason to believe that the general population in developed countries has greatly lessened physical activity in the last 100 years compared to earlier. Further, AD prevalence in developed countries is not markedly different in those segments of the population with considerable physical activity as opposed to those who are more sedentary.

So, based on the analyses of Chapters 4 and 5, copper and meat eating, the latter because it increases copper levels, are identified as environmental factors causative of the AD epidemic in the last century, in developed countries. Copper was originally suspect because first, it is present in increased amounts in amyloid plaques, second, when it binds to plaques it triggers release of damaging oxidant radicals, and third, it is known to be able to cause aggregation of beta amyloid into plaques. But a major tipping point came in 2003, when Sparks and Schreurs[1] showed that tiny amounts of copper, 0.12 ppm, when added to drinking water of an AD rabbit model, greatly enhanced amyloid plaque formation and memory loss in the animals. Copper is unique in this regard. It is the only putative AD causative agent that has actually been shown to cause amyloid plaque formation.

The evidence for copper as potentially AD causative was farther enhanced by a 2006 paper by Morris et al.,[2] which showed that ingestion of copper supplement pills, along with a high-fat diet, caused cognition loss at six times the rate of other groups.

Brewer[3] realized that the copper in the drinking water of the rabbits in the AD animal model study was inorganic copper and that it was greatly more toxic than the organic copper of food. The food copper could be increased 25 times more than 0.12 ppm of the drinking water, and not be toxic. Further, the copper in the supplement pills causing cognition loss in the Morris et al.[2] studies was also inorganic copper. Brewer's group[4] had earlier shown that some of the inorganic copper labeled with 64-copper was quickly (1–2 h) absorbed into the blood, while if the organic copper of food is radiolabeled, radiolabel does not show up in the blood for one or two days. The reason for the latter is that absorbed organic copper goes to the liver, where it is put into safe channels. Some of it is bound to proteins such as ceruloplasmin and secreted into the blood after one or two days. It was clear from the 64-copper studies that some inorganic copper was absorbed differently than organic copper, bypassing the liver, and much more quickly appearing in the blood. It was postulated by Brewer's group[3,5] that inorganic copper was AD causative.

The situation was further clarified by studies of Ceko et al.[6] They showed that food copper was mostly monovalent copper, or copper-1. This was surprising, since copper in living tissue, both plant and animal, is present in a redox doublet, both copper-1 and copper-2, because this allows the catalysis of many reactions critical to life. But it appears that at death, or harvest, in the absence of oxygen transport, copper-2 is reduced to copper-1. This leads to the explanation of the absorption of some inorganic copper (copper-2) quickly into the blood seen in the 64-copper studies. Because humans evolved ingesting primarily the copper-1 of food, an absorption system for copper-1, but not copper-2 has evolved. There is a copper absorption receptor, Ctr1,[7] in the small intestine that will absorb copper-1, but not copper-2. This receptor hands off the copper to a system that takes it to the liver, where it is put into safe channels. It does not appear in the blood for 2 days or so, and when it appears it is covalently bound to proteins, such as ceruloplasmin. Copper-2 cannot be absorbed by Ctrl. It can be absorbed by other routes, including diffusion, and some of it bypasses the liver, and appears immediately in the blood, where it can be toxic to cognition. Brewer[5] now refers to his inorganic copper hypothesis as the divalent copper hypothesis, or the copper-2 hypothesis, but they are the same hypothesis.

Copper-2 passes the two tests of causation of the AD epidemic in developed countries. First, it is directly tied to the pathogenesis of AD as shown clearly by the work of Sparks and Schreurs,[1] confirmed in additional studies by Sparks et al.,[8] and later confirmed in another laboratory in which tiny amounts of copper-2 in the drinking water of animal models greatly enhanced amyloid plaque formation in the brains of the animals and caused loss of memory.[9] Second, there has been greatly increased exposure to copper-2 in developed countries in the last 100 years, from two sources. The first of these is increased use of copper supplement pills in developed countries. Almost all multimineral preparations, many of which are paired with multivitamin preparations and are very popular in developed countries, contain copper as part of the supplement. This copper is all copper-2. The second source of copper-2 is drinking water. It has been shown that up to two-thirds of drinking water samples from homes in the United States contain enough copper to cause AD, if the animal model studies are a good guide.[10] The AD epidemic closely parallels the increasing use of copper plumbing in developed countries. Copper began to be used for plumbing in homes in the early 1900s, was curtailed by two world wars, and exploded after 1950, so that now, over 80% of US homes have copper plumbing. This is very similar to the pattern of the AD epidemic. The copper in drinking water is, of course, copper-2.

The above analysis indicts increased copper-2 ingestion as an agent causative of the AD epidemic. But there is also evidence indicting increased ingestion of copper in general as possibly partially causative of the AD epidemic. The Squitti group in Italy has shown that AD patients have an increased level of free copper in the blood.[11] Increased free copper in the blood is potentially toxic, as shown by its toxicity in the inherited disease of copper toxicity, Wilson's disease.[12] Further, the Squitti group has shown that the free copper levels in AD are negatively correlated with measures of cognition,[13] that higher levels of free copper are predictive of a higher rate of cognition loss,[14] and higher levels of free copper predict a higher probability of conversion of mild cognitively impaired (MCI) patients to full AD.[15] Thus, elevated free copper levels are intimately tied to the pathogenesis of AD which results in cognition loss. These data are highly suggestive that elevated free copper is causative of cognition loss in AD.

The Sqitti group has also found evidence of copper involvement of another type. They find an increased prevalence of ATP7B mutant alleles in the AD population.[16] ATP7B is the Wilson's disease gene causing, in patients who are homozygous for mutants that cripple this gene, copper accumulation and copper toxicity. Wilson's disease is recessive, meaning both copies of the gene have to be mutated to cause the disease. Patients who are heterozygous, meaning they have one copy of a mutated gene, have an increased copper load, in liver and urine for example, but do not require anticopper drug treatment. Assuming the ATP7B variants that the Squitti group find are increased in prevalence in AD patients also increase body copper loading, this indicates that mild increases in body copper loading over a lifetime, irrespective of the valence of the copper, is a risk factor for AD. This raises the question, why is homozygosity for Wilson's disease, which causes a huge increase in body copper loading and a huge increase in serum free copper, not causative of AD but rather produces a neurological movement disorder and no permanent effect on cognition and no amyloid plaques? The answer may be that homozygotes die young or are treated while young, while heterozygotes produce a mild increase in body copper loading for a lifetime that interacts with age, a strong AD risk factor, to produce AD.

The conclusion that mild increases in body copper loading over a lifetime is a risk factor for AD leads to the question, can this risk factor possibly be involved as partially causative of the AD epidemic? It would not be expected that the ATP7B alleles would be a factor in the AD epidemic, because these alleles are relatively rare and would not be expected to change much in frequency over the last 100 years. But finding that increased copper loading in general can be an AD risk factor raises the question, is there any other reason to cause mild increase in body copper loading in general? The answer is yes! As we have already discussed, there has been a major increase in meat eating in the last century in developed countries because of the great increase in raising animals in large quantities to supply meat for eating. As also previously pointed out, copper is much better absorbed from meat than from vegetable foods.[17] It is estimated that copper absorption is increased on an average of about 50% from a meat-containing diet as compared to a diet consisting of minimal meat and mostly vegetable foods. From this analysis, it is concluded that increased meat eating in the last 100 years in developed countries has contributed to the AD epidemic because of causing increased copper absorption and mildly increased body copper loading for a lifetime.

In conclusion, the search for culprits in causing the AD epidemic in developed countries in the last 100 years has yielded two, copper-2 specifically and increased body loading of copper in general irrespective of valence. The second mechanism, that is increased body loading of copper in general irrespective of valence, has been primarily facilitated by increased meat eating in developed countries in the last century.

REFERENCES

1. Sparks DL, Schreurs BG. Trace amounts of copper in water induce beta-amyloid plaques and learning deficits in a rabbit model of Alzheimer's disease. *Proc Natl Acad Sci USA* 2003;**100**:11065–9.
2. Morris MC, Evans DA, Tangney CC, Bienias JL. Dietary copper and high saturated and trans fat intakes associated with cognitive decline. *Arch Neurol* 2006;**63**:1085–8.
3. Brewer GJ. Copper toxicity in Alzheimer's disease. Cognitive loss from ingestion of inorganic copper. *J Trace Elem Med Biol* 2012;**26**:89–92.
4. Hill GM, Brewer GJ, Juni JE, Prasad AS. Treatment of Wilson's disease with zinc. II. Validation of oral 64 copper with copper balance. *Am J Med Sci* 1986;**292**:344–9.
5. Brewer GJ. Divalent copper as a major triggering agent in Alzheimer's disease. *J Alzheimers Dis* 2015;**46**:593–604.
6. Ceko MJ, Aitken JB, Harris HH. Speciation of copper in range of food types by X-ray absorption spectroscopy. *Food Chem* 2014;**164**:50–4.
7. Ohink H, Thiele DJ. How copper traverses cellular membranes through the copper transporter 1, Ctrl. *Ann NY Acad Sci* 2014;**1314**:32–41.
8. Sparks DL, Friedland R, Petanceska S, et al. Trace copper levels in drinking water, but not zinc or aluminum, influence CNS Alzheimer-like pathology. *J Nutr Health Aging* 2006;**10**:247–54.
9. Singh I, Sagare AP, Coma M, Perlmutter D. Low levels of copper disrupt brain amyloid-beta homeostasis by altering its production and clearance. *Proc Natl Acad Sci USA* 2013;**110**:14471–6.
10. Brewer GJ. Copper excess, zinc deficiency, and cognition loss in Alzheimer's disease. *Biofactors* 2012;**38**:107–13.
11. Squitti R, Pasqualetti P, Dal Forno G, et al. Excess of serum copper not related to ceruloplasmin in Alzheimer disease. *Neurology* 2006;**64**:1040–6.
12. Brewer GJ. Wilson's disease. In: Kasper DL, As F, Hauser SL, Longo DL, Jameson JL, Loscalzo J, editors. *Harrison's principles of internal medicine*. 19th ed. New York: McGraw-Hill Companies, Inc.; 2015.

13. Squitti R, Barbati G, Rossi L, Ventriglia M. Excess of nonceruloplasmin serum copper in AD correlates with MMSE, CSF [beta]-amyloid, and h-tau. *Neurology* 2006;**67**:76–82.
14. Squitti R, Bressi F, Pasqualetti P, Bonomini C. Longitudinal prognostic value of serum "free" copper in patients with Alzheimer disease. *Neurology* 2009;**72**:50–5.
15. Squitti R, Ghidoni R, Siotto M, Ventriglia M. Value of serum nonceruloplasmin copper for prediction of mild cognitive impairment conversion to Alzheimer disease. *Ann Neurol* 2014;**75**:574–80.
16. Squitti R, Polimanti R, Siotto M, Bucossi S. ATP7B variants as modulators of copper dyshomeostasis in Alzheimer's disease. *Neuromol Med* 2013;**15**:515–22.
17. Brewer GJ, Yuzbasiyan-Gurkan V, Dick R, Wang Y. Does a vegetarian diet control Wilson's disease? *J Am Coll Nutr* 1993;**5**:527–30.

CHAPTER SEVEN

Background on Copper, Including Why Copper-2 Is So Specifically Neurotoxic

INTRODUCTION

Much of this book will talk about the toxicity of copper. So while the toxicity of copper is important in the context of Alzheimer's disease (AD), it is important to understand that there is another side to copper—the good side. It is an essential element, which means that life is not possible without it. Not only that, but life becomes rather miserable if there is too great a lack of it, called copper deficiency. So this chapter will deal mostly with the good side of copper. The last section will deal with the neurotoxicity of copper-2. A very good review of the many roles of copper in metabolism will be found in the chapter on copper in the book written by Harris.[1]

COPPER AS A REDOX AGENT

As oxygen reached a higher concentration in the atmosphere, organisms evolved the capability to use it. Copper and iron became important in the utilization of oxygen because of their capability to easily gain or lose electrons, called reduction and oxidation, respectively. For short, they are called redox agents.

A very important reaction for metabolism in all higher animals is cytochromic oxidase. This enzyme contains both copper and iron and reduces molecular oxygen to water. The high-energy phosphate compound, adenosine triphosphate (ATP), is generated in the process. This kind of reaction allows organisms like the human to live in the fast lane, generating great energy from food and oxygen allowing organs like the brain and muscles that require a lot of energy, to function very effectively. Of course, this also comes at a cost. Redox activity generates as a by-product many reactive oxygen species (ROS) that can cause oxidant damage. Evolution has developed protective systems to scavenge and inactivate ROS, but if ROS

Environmental Causes and Prevention Measures for Alzheimer's Disease
ISBN 978-0-12-811162-8
http://dx.doi.org/10.1016/B978-0-12-811162-8.00007-X

generation is too great, some of the ROS escape inactivation and produce oxidant damage. In fact, one of the theories of aging is that it is due to continual oxidant damage.[2] If either copper or iron is present in excess, it causes toxicity through oxidant damage.

Copper is generally bound to proteins primarily by binding to nitrogen, such as in histidine, and to sulfur, such as in cysteine. Importantly, changing valence states during redox reactions occurs without disruption to the structure of the protein-binding site. This allows the copper to donate or receive electrons while still firmly bound to the protein, a critical feature for redox activity by the protein.

THE ESSENTIALITY OF COPPER FOR LIFE

Copper is an essential part of the catalytic function of many enzymes critical to life. One of these is lysyl oxidase, which is an enzyme important in the cross-linking of collagen and elastin connective tissue important in tendons, ligaments, blood vessels, and many other tissues. Thus, in copper deficiency, for example early in life, there are tendon and ligament abnormalities and tendencies for blood vessel aneurysms and rupture. The genetic disease, Menke's disease, which leads to copper deficiency, is a fatal disease early in life, emphasizing how important and essential copper is to life. More on this will be said in the sections in this chapter on copper deficiency and genetic copper diseases.

COPPER BALANCE IN HUMANS

The best data indicates that the average diet in developed countries contains about 1.0 mg of copper. This was the approximate average intake in 60 copper balance studies in which patients self-selected their diet.[3] Some authorities have set the daily requirement higher than that, but this is probably attributable to evaluating copper intake based on somewhat inaccurate tables of copper content of foods from older textbooks. The actual copper requirement in adults is about 0.5 mg/day. This is based on the average positive copper balance in untreated Wilson's disease (WD) patients of about 0.5 mg/day on an intake of 1.0 mg of copper/day.[3]

There is a little copper loss from the skin and a little copper loss in the urine, but in normal people by far the greatest copper loss is in the stool. And the mechanism of maintaining neutral copper balance is through excretion of copper in the stool. The liver accomplishes this by sensing whether copper levels are high or low. If high, an enzyme called ATP7B in the liver cell repositions itself and causes increased excretion of copper into the bile which is excreted into the bowel, and increased copper is lost in the stool. In this way copper balance is maintained. More copper is excreted into the bile if copper levels are high, and less if copper levels tend to be low. The balance responsibility of the liver for copper is unusual. Neutral balance of most cations is maintained by some combination of bowel absorption and urinary excretion.

Mutations in both copies of the ATP7B gene leads to WD, a disease in which copper levels increase and copper toxicity develops.[4] The mechanism, of course, is that with the ATP7B gene product crippled, the liver is no longer able to excrete the extra 0.5 mg copper/day taken in.

WD also helps teach about the relative absorption of copper from a vegetarian diet compared to a normal meat containing diet. It has been shown that a vegetarian diet controls WD.[5] Thus, the average 0.50 mg/day of positive copper balance typically seen in untreated WD disappears on a vegetarian diet. This means that absorption of copper is about 50% greater from a meat-containing diet than from a vegetarian one.

The bottom line for copper balance control in the human is that ATP7B of the liver is the sensor, probably sensing free copper levels and causing the excretion of more copper in the bile for loss in the stool as copper levels increase.

COPPER DEFICIENCY

There are several stages of copper deficiency, and each can be characterized by measuring the blood ceruloplasmin (Cp) levels. Cp is a protein synthesized by the liver and contains six or seven copper molecules. The amount of holo-Cp, that is Cp containing its copper molecules in the blood, is determined by copper availability in the liver. If copper levels begin to be low, the liver will continue to secrete apo-Cp, that is Cp without its copper molecules, into the blood, but the apo-Cp is rapidly cleared from the blood. Thus, the measurable Cp in the blood reflects the body copper status.

The first stage of copper deficiency is called chemical copper deficiency, because there is no clinical problem. Chemical copper deficiency has a medical use.[6] The Cp level, which is normally 18–35 mg/dL, can be lowered to 8–15 mg/dL by the drug tetrathiomolybdate (TM), a copper-lowering drug developed for WD, for medical purposes. At this level of copper depletion there are no clinical problems, but certain copper-dependent molecules are inhibited. This includes some angiogenic promoters (blood vessel growth-stimulating molecules). Since cancers need angiogenesis to grow, TM has anticancer effects through its antiangiogenic properties. Also there are many cytokines (biological signaling molecules) that are copper dependent. Inflammatory, fibrotic, and autoimmune processes are dependent on these cytokines. Thus TM has been shown to be effective against inflammatory, fibrotic, and autoimmune disease processes.[7]

The next stage of copper deficiency is mild clinical copper deficiency with a Cp of 2–7 mg/dL. In this stage there is mild anemia because copper is required for hemoglobin synthesis and mild leukopenia (low white blood cell count) because copper is required for cellular proliferation.

The final stage of copper deficiency is severe clinical copper deficiency with a Cp of 0–2 mg/dL. Anemia and leukopenia are present, and a neurological syndrome called myelopolyneuropathy appears.[8] The sensory nervous system is primarily involved, with minor involvement of the motor (muscular innervation) system. Patients develop numbness and paresthesias in the lower extremities first, which then spreads to the upper extremities. Loss of that part of the sensory system called proprioception, which senses where various parts of the body are, makes keeping balance difficult, and patients often lose the ability to walk. In addition to involvement of peripheral nerves, there is some involvement of nerve tracts in the spinal cord producing motor deficits. This severe disease does not have a good response to copper replacement therapy. The anemia and leukopenia is responsive to copper therapy, but usually the neurological disease is irreversible.

There are several causes of clinical copper deficiency. One of these is malabsorption of copper. Since copper is primarily absorbed from the small intestine, diseases of this organ, such as regional enteritis (Crohn's disease), or surgical removal of a significant part of the small intestine, can produce clinical copper deficiency. Gastric bypass surgery is sometimes causative of copper deficiency. Celiac disease (sprue) can also cause copper deficiency. Since a vegetarian diet results in less absorption of copper than a meat-containing diet, vegetarians are a little more susceptible to copper deficiency if some other causative factor is also present. These various copper malabsorption situations usually only produce mild copper deficiency (no neurological symptoms) and are usually completely responsive to copper supplementation. If the small intestine is not adequately capable of absorbing orally given supplementary copper, the copper can be given by parenteral (nonoral) routes.

Zinc therapy, or taking zinc supplements, can cause copper deficiency. Zinc induces intestinal cell metallothionein (MT), which binds copper in the intestinal cell, preventing its transfer to blood. The complexed copper is lost in the stool as the intestinal cells are sloughed, with about a 6 day lifespan. Whether zinc causes copper deficiency is a matter of zinc dose and whether zinc is taken 1–2 h away from food. Zinc taken close to food binds to substances in food and is poorly absorbed. So the following discussion of dose assumes the zinc is taken away from food. From considerable experience,[9] it is clear that doses of zinc up to 50 mg once daily are safe in adults. Studies in WD[3] revealed that it took 50 mg twice daily or 25 mg thrice daily to produce a negative copper balance, which, of course, is to be avoided in normal people trying to avoid copper deficiency. Bottom line—no more than 50 mg of zinc once daily should be taken.

Over the last 20 years there has been a great increase in cases of myelopolyneuropathy due to zinc-induced severe copper deficiency. It turns out that the cause was excessive use of dental adhesive containing high concentrations of zinc.[10,11] Some patients with ill-fitting dentures were using the adhesive repeatedly during the day, and swallowing large amounts of it, thereby receiving zinc doses of 200–300 mg/day. These patients had high serum zinc, absent Cp, anemia, leukopenia, and myelopolyneuropathy. Stopping use of the adhesive normalized the serum zinc and giving copper supplements normalized the anemia and leukopenia, but the neurologic symptoms were irreversible.

A major issue is whether a significant portion of the general population is mildly copper deficient and should receive copper supplements. There are a few scientists who claim that this is the case and that copper supplementation in the general population is beneficial. This is the rationale for multimineral supplement makers to include copper in their supplement pills. Probably half of the adult population in the United States takes a multivitamin/multimineral or multimineral supplement pill, and they all contain 0.5–1.0 mg of copper. But the evidence that a significant segment of the population has a copper deficiency problem is nonexistent. Hundreds to thousands of people have had Cp measurements, and there are no unexplained low values. It has been argued that Cp levels are not an adequate indicator of low copper status, because Cp is an acute phase reactant and is greatly elevated in situations such as inflammation and cancer, in which an acute phase response occurs. However, this argument is invalid. As discussed earlier[6,7] where TM was used to produce chemical copper deficiency to treat cancer, all these patients had elevated Cp levels to begin with due to an acute phase reaction, but as TM treatment ensued, Cp levels came down and became a reliable indicator of copper status.

So bottom line is that the general population does not need copper supplementation. And as shown by Morris et al.,[12] ingestion of copper supplements is terribly toxic to cognition and, as indicated in this book, is partially causative of AD. Copper supplementation to the general population should stop. Clinical copper deficiency is easily identified by Cp levels and the presence of anemia. These patients should be supplemented, and there is, now, supplement pills containing copper-1, rather than the toxic copper-2 which regular supplement pills contain.

Another point to be made from this discussion of clinical copper deficiency relates to the essentiality of copper, discussed in a prior section. As one looks at all the abnormalities in patients with severe clinical copper deficiency, it drives home how important adequate copper levels are to many critical parts of the body. That lesson will be driven home further in the next section, when Menke's disease, a disease of genetic copper deficiency, that is copper deficiency at birth, is discussed.

GENETIC COPPER DISEASES
Wilson's Disease

WD is due to inherited disabling mutations of both copies of the ATP7B gene,[4] which has been alluded to often in this book, most recently in discussing copper balance in a previous section of this chapter. As discussed in that section, ATP7B is the sensor in the liver of copper status in the body. If there is too much copper, it causes excretion of excess copper in the bile for loss in the stool.

Thus, when this sensor is disabled, as it is in WD, the small amount of excess copper taken in every day, estimated to be about 0.5 mg daily in the adult diet, accumulates. This excess copper is at first stored in the liver. After some years it damages the liver, and the patient may present with hepatitis (inflammation of the liver). If undiagnosed the hepatitis may abate, only to be repeated later. With continued liver damage, the patient may present with one of the complications of cirrhosis, which is severe fibrosis of the liver. Due to the fibrosis, pressure may increase in the portal vein, which normally delivers nutrients from the gastrointestinal system to the liver. The increased pressure in tributaries of the portal vein along the esophagus and stomach may lead to bleeding from these veins into the gastrointestinal tract and vomiting of blood. The increased pressure also leads to splenomegaly (enlarged spleen), which in turn can cause leukopenia (low white blood cell count) and/or thrombocytopenia (low platelet count) because the spleen is the body's filter for these cells. Finally, the liver disease of WD can present as acute liver failure, with very low serum albumin, edema (fluid accumulation), severe jaundice, and failure to synthesize adequate blood clotting factors. About half of WD patients present with some form of liver disease, usually between ages 15 and 25, although occasionally patients present with liver disease much later in life.

In about half of WD patients the liver is damaged but the patient does not present clinically. The liver is unable to store all the excess copper and the serum free copper increases. The next most sensitive organ to copper toxicity is the brain. Areas of the brain that control movement, such as the basal ganglia, are damaged, and the patient presents with a neurological movement disorder,[13] usually between the ages of about 18–30, although occasionally patients present much later in life. Symptoms may include tremor, dysarthria (difficulty with speech), dysphagia (difficulty with swallowing), and dystonia (rigidity of some muscles). The disease can progress until the patient is unable to walk or stand, and limbs may be pulled into grotesque positions by the dystonia.

Good drugs have been developed to treat WD, and if diagnosed early, further progression can be halted completely with effective therapy.[14] Recovery of existing symptoms is also usually relatively good with treatment, but there is always residual liver damage, and in neurological patients, often some residual symptoms. Further discussion of WD therapy, and discussion of diagnostic methods, is beyond the scope of the book, but there are many good reviews and books on the topic for those who are interested.

WD is an autosomal recessive disorder, meaning that both copies of the ATP7B gene have to be disabled to produce the disease. It has turned out that there are several hundred mutations of ATP7B discovered around the world that can produce the disease. Thus, often patients are compound heterozygotes, meaning that their two ATP7B genes have different disabling mutations. So far, it does not appear that phenotypic variation in age of onset or type of disease (hepatic vs. neurologic) can be explained by different ATP7B mutations. Rather it appears that variation in genetic background, and perhaps the environment, is the explanation for phenotypic variation. Patients who have one copy of an ATP7B mutation, but the second copy is normal, called heterozygous carriers, have a little increase in copper body load, manifested by a little increase in urine copper and liver copper, but do not require treatment.

In relevance to the main topic of this book, copper toxicity in AD, it is interesting that both AD and WD are diseases of copper toxicity involving the brain, with increased serum free copper levels, but are very different clinically and pathologically. AD is a disease of cognition loss, with no effect on movement control centers of the brain, and produces amyloid plaques and neurofibrillary tangles. WD is a disease of damaged movement control, no effect on cognition, damaged areas of the brain involved in movement control, but no amyloid plaques or neurofibrillary tangles. To add further to the puzzle, the Squitti group has found an increased frequency of carriers of ATP7B mutant alleles in the AD population.[15] If these mutants cause a mild increase in body copper load, it suggests that such a mild increase in body copper load and perhaps a mild increase in serum free copper is a risk factor for AD. Yet homozygous WD with its great increase in body copper load and serum free copper does not produce AD. The answer may be that to produce AD, the increase in body copper loading must work in conjunction with aging, a major risk factor for AD. Homozygotes for WD die young or are treated at a young age to normalize copper levels. If the WD is untreated, the basal ganglia are the first responders to the rather massive copper overload. Carriers of one copy of an ATP7B mutation have a mild copper overload that works in conjunction with aging to produce AD.

Menke's Disease

Menke's disease is due to a mutation in the ATP7A gene. Kaler[16] has written a very good review of Menke's disease and the other two ATP7A-related diseases. As the name suggests, the ATP7A gene bears close homology to the ATP7B gene. And the ATPase enzymes that both genes produce have a similar function, in that they both pump copper. But there are important differences. The ATP7B gene is expressed strongly in the liver where its product regulates copper balance, as previously discussed, and is expressed almost nowhere else. The ATP7A gene is broadly expressed, thus its product serves as a copper pump in many locations. The loss of this function in many sites produces the numerous defects and symptoms of Menke's disease, to be discussed shortly. Another difference is that while the ATP7B gene is on an autosome, the ATP7A gene is located on the X-chromosome, so defects in it produces sex-linked disorders, primarily in males, who have only one X-chromosome.

Menke's disease is a disease of infantile onset with failure to thrive, coarse kinky hair, neurodegeneration, connective tissue abnormalities, and other defects.[16] A key role for the ATP7A enzyme is to pump copper out of the small intestinal cell for absorption of copper, so these patients become severely copper deficient after birth, although they are born not copper deficient, because apparently the mother's placenta supplies adequate copper while the fetus is in the uterus. The ATP7A enzyme also is important in getting copper from the blood into the brain, so the brain fails to develop in these babies, and they are severely mentally retarded.

As part of the connective tissue abnormalities, the blood vessels, particularly of the brain, become tortuous and tend to rupture. Also the hair becomes kinky and there are many other connective tissue abnormalities. The connective tissue abnormalities are due to a lack of activity of an enzyme called lysyl oxidase, discussed earlier, which is a copper-containing enzyme that requires the copper for activity.

These babies tend to die at about age 2–3 years. Some live longer because they have a mutation that has not completely inactivated the ATP7A gene. Treatment is with parenteral copper administration, which is only effective if the particular mutation results in at least 10% of normal ATP7A enzyme activity.[16]

Occipital Horn Syndrome

Occipital horn syndrome (OHS) is also due to mutation in the ATP7A gene, but results in a milder disease than Menke's disease, because the mutations allow retention of 20%–30% activity.[16] The syndrome gets its name from the horn-shaped calcifications that occur bilaterally within the tendons of muscles attached to the occipital bone, which can be seen on X-rays.

These patients share the abnormal hair and some of the connective tissue abnormalities of Menke's patients. The neurological abnormalities of OHS are mild and do not become apparent until late childhood. They are primarily due to a deficiency of a copper-dependent enzyme, dopamine-B-hydroxylase, which results in a generalized mild muscle weakness and dysautonomia which produces a tendency to fainting, orthostatic hypotension, and diarrhea. The long-term outlook for these patients is uncertain because not many have been followed long enough to develop a clear picture. Treatment is still experimental.

ATP7A–Related Distal Motor Neuropathy

This is a newly recognized ATP7A-caused disease, resembling Charcot-Marie-Tooth disease. The disease begins at age 10–35 years and produces a distal motor neuropathy. It begins with weakness and atrophy of the legs then usually moves to the arms. Patients often develop foot and hand deformities.[16]

The type of ATP7A mutation that produces the disease appears to interfere with intracellular trafficking of the ATP7A enzyme. This type of mutation appears to be specific for peripheral nerves. In this disease there are none of the defects of Menke's disease or OHS, and those diseases do not have neuropathy.

In the summary of the genetic copper diseases, WD teaches about the toxicity of copper, very relevant to the general theme of this book, copper toxicity in AD. The ATP7A genetic diseases teaches strongly about the essentiality of copper, namely the very large number of things that go wrong if copper is lacking. Literally, copper is essential for life.

COPPER-2 VERSUS COPPER-1: WHY COPPER-2 IS SO SPECIFICALLY NEUROTOXIC

The final piece of "Background on Copper," which is what this chapter is about, is the critical difference between copper-2 and copper-1, a major theme of the book in terms of AD causation. As stated before, copper is a redox-active element meaning it can easily go back and forth between the oxidized Cu^{++} or copper-2 state, and the reduced Cu^+ or copper-1 state. This redox capability has been taken advantage of in evolution, with the development of many redox reactions critical to life dependent on copper.

Copper is a very toxic element, so it is important that copper ions always be bound to molecules such as proteins. Copper must be transported through multiple steps where it is transferred from one protein to another, and because of the toxicity of free ionic copper, evolution has developed "chaperone" molecules to bind copper and move it from one protein to another.[1]

It has become clear that copper in food is primarily copper-1.[17] While copper-2 is also present in tissues, because it is known that copper-2 and copper-1 form a redox doublet allowing reactions critical to life, it appears that at death or harvest, in the absence of oxygen transport, copper-2 becomes reduced to copper-1. Thus, because human ancestral organisms, and humans themselves, ingested primarily copper-1 and not much copper-2, the careful, safe, step-by-step handling of copper during absorption from the gastrointestinal tract only evolved for copper-1. Thus, there is a copper-1-specific receptor in the intestinal cell, Ctr1,[18] which moves copper-1 into the intestinal cell where it enters pathways destined to transport it to the liver, where it continues its journey, step-by-step, always chaperoned, until it ends up with target proteins in the liver cell.

Ctr1 does not recognize copper-2, so it is not absorbed through this safe pathway, unless it is reduced to copper-1. Copper-2 can be absorbed by diffusion and by the divalent cation transporter. We know that unlike copper-1, which when radiolabeled does not appear in the blood for 1 or 2 days and is then safely incorporated into a protein secreted by the liver, a substantial portion of ingested copper-2, at least about 25%, appears in the blood 1–2 h after ingestion.[19] This is much too fast to be processed by the liver. This copper-2 appears to add immediately to the free copper pool of the blood and is toxic to cognition.

So, along comes the 20th century and what happens? For the first time, humans in developed countries are exposed to ingestion of copper-2 from drinking water, in relation to the "epidemic of copper plumbing," primarily in the last half of the century. It has been shown that enough copper leaches from the copper plumbing into the water to cause AD, if the animal models are a good guide.[20] And during this same century, primarily in the last half of the century, an "epidemic" of AD occurs in developed countries. Adding to the copper-2 ingestion in the 20th century in developed countries is the increasingly common habit of taking a multimineral pill containing copper-2.

So is the very close time concurrence of the two "epidemics," the copper-2 ingestion one and the AD one, just coincidence? After all, many things change during development. In Chapters 4 and 5 considerable effort was spent on examining other possibilities, and none other than copper-2 and copper in general were found to fit. In the next chapter, Chapter 8, the total web of evidence supporting the critical role of copper-2 as causing the AD epidemic in developed countries will be presented. That web of evidence should convince all that the time concurrence of the copper plumbing and AD epidemics is not coincidence, but a cause and effect relationship. And in the following chapter, Chapter 9, the additional role of a mildly increased body load of copper for a lifetime, resulting from increased meat ingestion, contributing to the AD epidemic will be examined.

REFERENCES

1. Harris ED. *Minerals in food, nutrition, metabolism, bioactivity.* Lancaster (PA): DEStech Publications, Inc.; 2014.
2. Harman D. Aging: a theory based on free radical and radiation chemistry. *J Gerontol* 1956;**11**:298–300.
3. Brewer GJ, Yuzbasiyan-Gurkan V, Dick R. Zinc therapy of Wilson's disease VIII: dose response studies. *J Trace Elem Exp Med* 1990;**3**:227–34.
4. Brewer GJ. In: Kasper DL, Fauci AS, Hauser SL, Longo DL, Jameson JL, Loscalzo J, editors. *Wilson's disease in Harrison's principles of internal medicine.* 19th ed. New York: McGraw-Hill Co. Inc.; 2015.
5. Brewer GJ, Yuzbasiyan-Gurkan V, Dick R, Wang Y, Johnson V. Does a vegetarian diet control Wilson's disease? *J Am Coll Nutr* 1993;**12**:527–30.
6. Brewer GJ. The promise of copper lowering therapy with tetrathiomolybdate in the cure of cancer and in the treatment of inflammatory disease. *J Trace Elem Med Biol* 2014;**28**:372–8.
7. Brewer GJ. The use of copper-lowering therapy with tetrathiomolybdate in medicine. *Expert Opin Investig Drugs* 2009;**18**:89–97.
8. Kumar N, Grum B, Petersen RG, Vernino SA, Ahlskog JE. Copper deficiency myelopathy. *Arch Neurol* 2004;**61**:762–6.
9. Brewer GJ, Rd D, Johnson VD, Brunberg JA, Kluin KJ, Fink JK. The treatment of Wilson's disease with zinc: XV Long-term follow-up studies. *J Lab Clin Med* 1998;**132**:264–78.
10. Nations SP, Bower PJ, Love LA, et al. Denture cream: an unusual source of excess zinc, leading to hypocupremia and neurologic disease. *Neurology* 2008;**71**:634–9.
11. Hedera P, Peltier A, Fink JK, Wilcock S, London Z, Brewer GJ. Myelopolyneuropathy and pancytopenia due to copper deficiency and high zinc levels of unknown origin II. The denture cream is a primary source of excessive zinc. *Neurotoxicology* 2009;**30**:996–9.
12. Morris MC, Evans DA, Tangney CC, Bienias JL. Dietary copper and high saturated and trans fat intakes associated with cognitive decline. *Arch Neurol* 2006;**63**:1085–8.
13. Brewer GJ. Neurologically presenting Wilson's disease: epidemiology, pathophysiology, and treatment. *CNS Drugs* 2005;**19**:185–92.
14. Brewer GJ, Askari FK, Lorincz MT, et al. Diagnosis and treatment of Wilson's disease with an update of anticopper treatment for other diseases. *Biomed Res Trace Elem* 2004;**15**:211–21.
15. Squitti R, Polimanti R, Siotto M, Bucossi S. ATP7B variants as modulators of copper dyshomeostasis in Alzheimer's disease. *Neuromol Med* 2013;**15**:515–22.
16. Kaler SG. ATP7A-related copper transport diseases-emerging concepts and future trends. *Nat Rev Neurol* 2011;**7**:15–29.
17. Ceko MJ, Aitken JB, Harris HH. Speciation of copper in range of food types by x-ray absorption spectroscopy. *Food Chem* 2014;**164**:50–4.
18. Ohink H, Thiele DJ. How copper traverses cellular membranes through the copper transporter 1, Ctrl. *Ann NY Acad Sci* 2014;**1314**:32–41.
19. Hill GM, Brewer GJ, Juni JE, Prasad AS, Dick RD. Treatment of Wilson's disease with zinc. II. Validation of oral 64 copper with copper balance. *Am J Med Sci* 1986;**292**:344–9.
20. Brewer GJ. Copper excess, zinc deficiency, and cognition loss in Alzheimer's disease. *Biofactors* 2012;**38**:107–13.

CHAPTER EIGHT

Inorganic Copper, or Copper-2, Ingestion as a Major Causal Factor for the Alzheimer's Disease Epidemic—The Web of Evidence

To prove that a particular agent causes a disease, it is desirable to fulfill Koch's criteria, the major one of which is to give the agent and produce the disease. In this chapter, strong evidence, approaching proof, is given that copper-2 ingestion is a major causal factor in the Alzheimer's disease (AD) epidemic in developed countries over the last century. It is unethical to do the strongest case trial in humans, that is, give some people copper-2 and others placebo, thus doing a randomized clinical trial (an RCT). However, what has already happened in humans comes close to an RCT, and RCTs have already been done in AD animal models, and are positive! After presenting the data in this chapter, these statements will be enlarged upon and clarified in the chapter summary.

If one cannot do an RCT, the next best thing is to develop the evidence for the assertion of causality, trying to build such a strong web of evidence that any reasonable person, having seen the evidence, would agree with the assertion of causality. In this chapter, such a web will be developed around copper-2 causality of AD.

AD ANIMAL MODEL STUDIES

The AD rabbit model studies of Sparks and Schreurs[1] was gone into in detail in Chapter 1. Recapitulating, they found that when they moved their laboratory from West Virginia to Arizona, they could not reproduce the positive results in their model. This was a cholesterol-fed model in rabbits, in which amyloid plaques in the brain and memory loss by the animals were the major features. They finally realized that a major difference was that they were using distilled drinking water in Arizona while they had used tap water for drinking water in West Virginia. It was clear that something in tap water was key to producing the AD disease. Investigations determined that trace amounts of copper (0.12 ppm) in the drinking water made

Environmental Causes and Prevention Measures for Alzheimer's Disease
ISBN 978-0-12-811162-8
http://dx.doi.org/10.1016/B978-0-12-811162-8.00008-1

all the difference. The presence of 0.12 ppm copper, which is copper-2, in the distilled drinking water produced amyloid plaques and memory loss. Distilled drinking water without copper did not produce amyloid plaques and memory loss. This is an RCT in the animal model, and it was strongly positive!

Later Sparks et al.[2] extended their studies to several AD animal models, many of which were not especially fed cholesterol or other substances, and duplicated their findings of 0.12 ppm copper causing enhanced plaque formation. They even tried aluminum and zinc, two other metals that some thought were AD causative and showed that only copper produced the AD-enhancing effect.

The findings of the Sparks group, that trace amounts of copper in the drinking water caused AD, were confirmed in another laboratory using a mouse AD model.[3]

In contrast to the effect of trace amounts of copper added to drinking water, changing the copper in the food of the animals by much larger amounts, for example, 3–6 ppm, a 25-fold greater change than in the drinking water copper, would have no effect on either amyloid plaques or on memory. What this shows is that copper in drinking water is exquisitely more toxic to the brain than copper in food. And, of course, copper in water is copper-2, while copper in food is copper-1. As pointed out earlier, humans evolved a system to safely absorb copper-1, but not copper-2. Copper-2 in the water is a poison to the brains of both animals and humans, and Sparks and Schreurs[1] deserve the credit for first pointing this out in their AD animal model studies.

STUDIES OF THE EFFECTS OF INGESTION OF INORGANIC COPPER IN SUPPLEMENT PILLS ON HUMAN COGNITION

Just as striking as the 2003 paper by Sparks and Schreurs,[1] but in a different way, was a 2006 paper by Morris and colleagues.[4] This group studied the intake of all kinds of nutrients in a large Chicago population, and looked at these peoples' cognition and cognition change over several years. They found that those in the highest one-fifth of copper intake, if they also ingested a high-fat diet, lost cognition at six times the rate of other groups! These people were in the highest one-fifth of copper intake because they were taking multimineral supplement pills containing copper-2.

So, if a person ingests a high-fat diet, as many Americans and also people in other developed countries do, and the person also takes a multimineral- or multivitamin/multimineral-containing copper, that person has a major risk of rapidly losing cognition. Since probably half the population takes one of these pills, and almost all contain copper, this appears to be a major public health risk.

Since the Morris et al.[4] paper came out in 2006, and since ingesting copper supplement pills seems to impose a major health risk, the US FDA, who has the responsibility to ensure the safety of such supplements, should have taken action by now. But they have not. It appears the population is left to its own devices to protect itself against copper-2 ingestion from both supplement pills and drinking water. This is an important, perhaps critical, reason for reading this book and spreading the message.

The threat of copper supplement pill ingestion is made even more dire by a recent study published by Mursa et al.[5] In this study, ingestion of various supplements was determined in a large number of women, and related to mortality. (In the interest of full disclosure, this is not mortality from AD. It is overall mortality but would of course include mortality related to AD.) They found that in women ingesting supplementary copper, there was a 42% increased risk of all-cause mortality (meaning death from all causes). The risk of dying in a given period of time is 42% higher if a copper pill is ingested! The Mursa et al.[5] study, published in 2011, is now several years old, but so far there is no indication that the FDA is taking action, not even finding a way to see if the Mursa et al.[5] study can be confirmed. So the supplement pill makers keep putting the copper poison in their pills and raking in the money.

STUDIES SHOWING THAT SOME INORGANIC COPPER, IN CONTRAST TO ORGANIC COPPER, IS ABSORBED DIRECTLY INTO THE BLOOD

After the Sparks and Schreurs[1] and Morris et al.[4] studies were published, the question arose, what do copper in drinking water and copper in supplement pills have in common? The answer, of course, is that they are both inorganic copper, or copper-2. This raises the question, is inorganic copper handled differently by the body than organic, or food copper? It appears the answer was already available, from studies done several years earlier.

To briefly recap these studies, they occurred when zinc was being developed as a therapy for Wilson's disease, approved by the FDA in 1997. Wilson's disease is a disease of copper accumulation and copper toxicity. Zinc acts by blocking copper absorption in the small intestine. A radioactive copper, copper-64, was used to test whether zinc blockade of copper absorption was effective.[6] An oral dose of copper-64 was given and blood radioactivity at 1 and 2 h after giving the oral dose of copper-64 was determined. Before zinc administration, there would be a large peak of radioactivity from copper-64 at 1 and 2 h. After effective zinc administration for 2 weeks, no blood radioactivity would appear at the 1 and 2 h time points after oral copper-64 administration, showing that zinc therapy was working.

The significance of these studies is as follows. Copper-64 was given as an inorganic salt of copper. In other words, it was a marker of the absorption of inorganic copper. If the copper of food copper was made radioactive, no radioactivity would appear in the blood for several days. Food, or organic, copper is transported to the liver after absorption from the intestine, where it is put into safe channels. These safe channels include incorporation into proteins that are made by the liver and secreted into the blood. It is these secreted proteins that contain the radio copper that cause blood radioactivity several days after giving labeled food copper.

But in the absence of zinc there is substantial blood radioactivity 1 and 2 h after oral administration of copper-64 as an inorganic salt. It is estimated that perhaps as much as 25% of the administered dose of copper-64 is showing up in the blood almost immediately. This is much too soon for the copper to have been processed by the liver. So, it is clear from this, that a substantial portion of ingested inorganic copper bypasses the liver and appears immediately in the blood, adding directly to the free copper pool of the blood.

Explaining the term free copper pool of the blood, blood copper can be divided into two pools. About 85% of blood copper is incorporated into the protein ceruloplasmin and is safe copper. The other 15% of blood copper is loosely bound to various molecules and is called free copper. This free copper is the copper readily available for the body's needs. However, it is the potentially toxic copper if the free copper pool becomes larger. For example, in Wilson's disease, the free copper pool becomes considerably enlarged, and this copper becomes very toxic. So it may be noteworthy that part of the ingested inorganic copper is added immediately to the free copper pool.

So, summarizing, it appears that the inorganic copper of drinking water and the inorganic copper of supplement pills, partially bypass the liver and are added immediately to the blood's free copper pool. Later in this chapter we will discuss in detail that inorganic copper is copper-2 (Cu^{++}, or divalent copper) while food, or organic, copper is copper-1 (Cu^+ or monovalent copper).[7] Humans evolved to safely handle food copper (copper-1). There is a specific copper intestinal transport system, called Ctr1,[8] for copper-1, which results in depositing the copper-1 into the liver for safe handling. But humans were not exposed to copper-2 during their evolution, and as a consequence, there is no safe absorption route for copper-2. Copper-2 cannot be absorbed by Ctr1 until it is reduced to copper-1.

Ingested copper-2 is absorbed by diffusion and the divalent cation transporter. Obviously from the copper-64 studies, some of this copper-2 bypasses the liver and appears in the blood free copper pool immediately. Equally obvious, this copper-2 eventually causes brain damage leading to loss of cognition and to AD.

THE EPIDEMIC OF ALZHEIMER'S DISEASE CORRELATES TIME-WISE WITH THE SPREAD OF COPPER PLUMBING

In Chapter 3 it was emphasized that the epidemic of AD was new, developing over the last century, with a major increase in prevalence over the last 50 years, with the epidemic predominant only in developed countries. It was asserted that these facts make it very obvious that a new environmental agent or agents, newly present in developed countries, was causative of the epidemic of AD. The studies of Sparks and Schreurs[1] made it clear that inorganic copper in drinking water could be causative. But is this relevant to humans? Is there copper present in human drinking water? One obvious source would be leaching of copper from copper plumbing. So, what is the history of copper plumbing, and how widely has it spread?

The increasing use of copper plumbing in developed countries after 1900[9] parallels remarkably the increasing prevalence of AD in these countries. Like the history of AD, copper plumbing began to be used in the early 1900s. But the use of copper was curtailed by World War I, and then again by World War II. AD was increasing slowly during this period. After 1950, the use of copper plumbing exploded, so that now over 80% of US homes have copper plumbing. During this period the prevalence of AD also exploded. Copper plumbing is not used very much in undeveloped countries because of the expense, and these countries have not shared in the AD epidemic.

Japan is an interesting case which supports the role of copper plumbing in causing the AD epidemic. It is a developed country with a low prevalence of AD.[10] And guess what? Japan has shunned copper plumbing, apparently for fear of toxicity! Yet when Japanese move to Hawaii, where copper plumbing is used, the prevalence of AD increases and is similar to other developed countries.[11]

So in summary, inorganic copper (copper-2) in the drinking water could be the environmental culprit causative of AD if sufficient copper is leached from copper plumbing to reach the levels shown to be toxic in the animal models of AD. The next section will be enlightening to see if that criterion is met.

STUDIES WHICH SHOW THAT SUFFICIENT COPPER IS LEACHED FROM COPPER PLUMBING INTO DRINKING WATER TO CAUSE AD

Again, the older Wilson's disease studies answer the question about the copper levels in human drinking water in developed countries. Because Wilson's disease is a disease of copper, it could be important to minimize the intake of copper in drinking water in Wilson's disease patients. So each patient sent in a sample of their drinking water in special trace element free tubes provided. They were instructed to avoid "first draw" water which had been stagnating in the plumbing pipes overnight.

The study collected 280 household drinking water samples from homes all over North America, since the Wilson's disease patients were from all over North America. Assays of the copper levels in these samples revealed that[12]:

1. Roughly one-third were 0.1 ppm or higher, the level found toxic in AD animal models.
2. Roughly one-third were at 0.01 or below, a level believed to be safe.
3. Roughly one-third were between 0.01 and 0.1 ppm, levels of unknown safety because they have not been studied in animal models.

So, the bottom line is that there is plenty of copper in one-third to two-thirds of the drinking water samples in North America to cause AD, if the animal models are a good guide, as they likely are.

Thus, the drinking water copper data, and the time-wise spread of copper plumbing, fit very well with inorganic copper (copper-2) in drinking water being a major environmental culprit in causing AD.

DATA SHOWING THAT COPPER IN FOOD IS MOSTLY COPPER-1

It was always assumed that copper in food is a mixture of copper-1 and copper-2. That is because in living plant or animal tissue, copper-1 and copper-2 form a redox doublet, helping to catalyze many reactions important to life. Then Ceko et al.[7] published a paper in which they determined the speciation of copper (that is, copper-2 vs. copper-1) by X-ray spectroscopy in a variety of foods as well as in drinking water.

They found that in drinking water copper was copper-2, as expected, but in foods, unexpectedly, copper was mostly copper-1. Apparently at death or harvest, in the absence of oxygen transport, copper-2 is reduced to copper-1. This is a very important finding. It means that animals and humans evolved ingesting primarily copper-1 and were not exposed much to ingesting copper-2 until the 20th century in developed countries.

HOW EVOLUTION SHAPED HUMANS FOR SAFE INGESTION OF COPPER-1 BUT NOT COPPER-2

Copper is a very toxic element, but essential for life. So during evolution, safe channels had to be developed to handle it. For example, there are numerous copper chaperones that safely bind copper and move it from one protein to another where it is needed. In the intestinal tract, when copper-1 comes in with food, there is a receptor, called Ctr1[8] which binds copper-1 and routes it to the ATP7A enzyme which puts it in a vesicle headed for the liver, where it is carefully routed to various safe channels, to carry out functions. However, Ctr1 does not bind copper-2, so it cannot be transported through this carefully evolved route. Copper-2 can be absorbed through diffusion or by the divalent cation transporter. It is clear that at least some bypasses the liver and appears in the blood, adding to the free copper pool, and over time, is toxic to cognition.

This difference in absorption routes for copper-1 and copper-2 explains the results discussed in an earlier section of this chapter, when it was pointed out that some inorganic copper labeled with 64-copper appears immediately in the blood,[6] while labeled organic copper of food shows labeled copper in the blood only after one or two days. Obviously, the inorganic 64-copper is copper-2 and some is bypassing the liver and appearing immediately in the blood. In contrast the labeled organic (food) copper is all passing through the liver, and its metabolism there and incorporation into proteins secreted by the liver into the blood takes one or two days.

STUDIES IN CHINA WHICH SHOW THAT AD PREVALENCE CORRELATES POSITIVELY WITH SOIL CONCENTRATIONS OF COPPER ACROSS THE PROVINCES OF CHINA

Shen et al.[13] have examined the relationship between AD prevalence and soil concentrations of copper across most of the provinces of mainland China. They find that there is a highly significant positive correlation. For example, the relative risk of AD is 2.6 times higher if the soil copper concentration is 60–80 ppm compared to a soil copper concentration of 20–40 ppm.

It appears likely that the drinking water concentration across these Chinese provinces would correlate with the soil copper concentration. If this is so, it fits with the mechanism for AD causation expounded on here. That is, in areas where there is a high concentration of copper in the soil, it is likely that there is also a high concentration of copper-2 in drinking water, and then a high prevalence of AD occurs in that area because of ingestion of a large amount of copper-2 in the drinking water.

Of course, this explanation for the correlation between soil copper concentration and AD prevalence remains somewhat speculative until drinking water copper levels are determined across these provinces. Until then it remains a likely supporting piece of data for the concept of copper-2 ingestion as causative of AD.

ANIMAL STUDIES OF THE GREATER TOXICITY OF COPPER IN DRINKING WATER THAN IN FOOD

In a recent study, Wu et al.[14] gave mice their entire copper intake in either drinking water or in food. Copper was given at 6, 15, and 30 ppm. The 6 ppm part of the study is looking at normal physiological intake. At all three levels of copper intake, toxicity of copper taken in via drinking water was greater than that of copper taken in in food. The toxicity was shown by lower liver levels of superoxide dismutase in the food copper animals, indicating greater oxidant stress, and increased levels of soluble beta amyloid in the brain, indicating a greater tendency toward AD. At the higher copper intakes, in the water copper animals, oxidant stress was increased, and serum free copper was elevated compared to the food copper animals.

It appears likely that in the above study the copper given in food was reduced to copper-1 and bound to amino acids and proteins by the time it reached the intestine, while copper in drinking water remained as copper-2. So this study shows first, that copper in drinking water is much more toxic than copper in food, second that copper-2 is much more toxic than copper-1, and third that the copper-2 in drinking water is "amyloidogenic," meaning that it is AD enhancing.

SUMMARY AND CONCLUSIONS

Table 8.1 lists the nine sets of data and observations that form the web of evidence supporting copper-2 as a major environmental causal factor of the AD epidemic in developed countries.

Reviewing the process used from the beginning, Chapter 3 presented strong, well-documented evidence that AD was rare in the 1800s, probably

Table 8.1 Listing of the Evidence That Copper-2 Is AD Causative

Point	Evidence	References
1	Tiny amounts of copper-2 in drinking water enhance AD pathology and inhibit memory in AD animal models.	1–3
2	Ingestion of copper-2 in copper supplement pills, in those who also ate a high-fat diet, reduces cognition in humans at a sixfold normal rate.	4
3	Some ingested copper-2 by humans is absorbed directly into the blood bypassing the liver, while all absorbed copper-1 passes through the liver.	6
4	The epidemic of AD in developed countries correlates time-wise with the spread of copper plumbing.	9
5	Studies show sufficient copper is leached from copper plumbing into drinking water to cause AD, using animal models as a guide.	12
6	Data shows that most of the copper in food is copper-1.	7
7	Evolution developed safe channels for copper-1 but not copper-2.	8
8	Studies in China show that AD prevalence correlates positively with soil copper concentrations across Chinese provinces.	13
9	Animal studies show that copper-2 in drinking water is much more toxic than equivalent amounts of copper in food, and is amyloidogenic, meaning AD favoring.	14

with a prevalence of about 1%, and then became more and more frequent in the first half of the 20th century, and then exploded to a high prevalence in the last half of the 20th century, but in developed countries only. Since AD is a disease of aging, the increasing lifespan of people in the 20th century had to be considered as a factor in increasing AD prevalence. Increasing lifespan certainly increases the overall caseload, but it is not an explanation for the very low prevalence in the 1800s, because data from both France and the United States show that there were plenty of older people around 1900, enough to have provided a large number of AD patients to show up in clinics and at the autopsy table at today's rate.

The sole remaining explanation for the AD epidemic in developed countries is that an environmental factor or factors emerged in the 20th century in developed countries that was causative of AD. In Chapters 4 and 5, a careful examination of candidate environmental agents was carried out, using two criteria. First, there must be supporting information that the putative agent could be tied to AD pathogenesis. Second, exposure to the agent had to be greatly increased in the 20th century compared to the 19th century. Out of this examination of possible new environmental AD causative agents emerged copper-2 and a lifetime of mildly increased body copper load. This chapter is focused on copper-2. In the next chapter the focus will be on increased body copper load.

If one wants to claim that an agent causes a disease, it is nice to fulfill Koch's postulates, the last one of which is to give the agent and cause the disease. Of course, it would be unethical to deliberately give copper-2 to humans to cause AD. So short of that, the next best thing is to develop a web of evidence that so consistently supports the agent as causative that any reasonable person, after examining the evidence, would be convinced of its causal role. That is what has been done here, and the web of evidence is summarized in Table 8.1.

Commenting on the specific pieces of evidence in Table 8.1, the first point is particularly powerful. It actually fulfills Koch's last postulate, but in animals, not in humans. Tiny amounts of copper-2 given in drinking water actually cause AD in the animals.[1–3] And it is actually a randomly controlled trial (an RCT), which is the highest level of evidence in human trials. Randomly selected animals were either given copper, or nothing (placebo), or aluminum (a control), or zinc (another control).[2] Only the copper was AD causative. These studies are as close as you can get to proof of copper-2 AD causation in animal models. For some reason its power has not been recognized by the AD scientific community.

The second point of Table 8.1 is also powerful. It actually fulfills Koch's last postulate, in that humans actually gave themselves the putative causative agent. And lo and behold—if they also ate a high-fat diet, they lost cognition at six times the rate of other groups.[4] While this study did not follow people until AD developed, it is not hard to believe that those rapidly losing cognition were often headed to AD.

Point 3 of Table 8.1 explains why copper-2 ingestion is more toxic than copper-1 ingestion. All absorbed copper-1 passes through the liver, where it is put into safe channels. Radiolabeled food copper, primarily copper-1, does not appear in the blood for a day or two, having passed through the liver and being incorporated into proteins secreted by the liver into the blood. In contrast, some absorbed copper-2 appears in the blood within an hour or two, as shown by 64-copper[6] (copper-2) radiolabeling, much too soon to be processed by the liver. This copper-2 adds to the free copper pool and is toxic to cognition.

Points 4 and 5 of Table 8.1 both deal with copper-2 in drinking water causing AD. Point 4 shows that the two epidemics, the AD epidemic and the "epidemic" of using copper plumbing in developed countries, were concurrent.[9] Both began early in the 20th century, gradually increased, and then both exploded after 1950. Point 5 makes point 4 significant, because it shows that enough copper is leached from copper plumbing to cause AD in a large proportion of cases.[12]

Point 6 of Table 8.1 is a bombshell! Surprisingly, despite the fact that in life copper forms a redox doublet, requiring the presence of both copper-2 and copper-1, point 6 reveals that food copper is mostly copper-1.[7] At death, or harvest, in the absence of oxygen transport, most of copper-2 is reduced to copper-1. This finding explains point 7 of Table 8.1. Since human ancestors and humans themselves were exposed primarily to ingesting copper-1 but not copper-2, evolution developed safe channels for handling only copper-1. In turn, point 7 explains point 3, namely the channels routing absorbed copper-1 and copper-2 are different, allowing some copper-2 direct access to the blood free copper pool. In turn, this explains the toxicity of copper-2, shown in points 1 and 2.

Point 8 of Table 8.1 [13] is further support for the copper-2 hypothesis, if the soil concentration of copper is positively correlated with the copper concentration of drinking water and, of course, drinking water copper would be copper-2.

Point 9 of Table 8.1 is further support for point 1, with a different type of animal experiment.[14]

In summary, the points raised in this chapter, and summarized in Table 8.1, are deemed to provide an overwhelming web of evidence in support of copper-2 ingestion as a major environmental factor causing the AD epidemic in the 20th century in developed countries.

REFERENCES

1. Sparks DL, Schreurs BG. Trace amounts of copper in water induce beta-amyloid plaques and learning deficits in a rabbit model of Alzheimer's disease. *Proc Natl Acad Sci USA* 2003;**100**:11065–9.
2. Sparks DL, Friedland R, Petanceska S, et al. Trace copper levels in drinking water, but not zinc or aluminum, influence CNS Alzheimer-like pathology. *J Nutr Health Aging* 2006;**10**:247–54.
3. Singh I, Sagare AP, Coma M, Perlmutter D. Low levels of copper disrupt brain amyloid-beta homeostasis by altering its production and clearance. *Proc Natl Acad Sci USA* 2013;**110**:14471–6.
4. Morris MC, Evans DA, Tangney CC, Bienias JL. Dietary copper and high saturated and trans fat intakes associated with cognitive decline. *Arch Neurol* 2006;**63**:1085–8.
5. Mursa J, Robien K, Hamack LJ, Park K, Jacobs DR. Dietary supplements and mortality rate in older women: the Iowa Women's Health Study. *Arch Int Med* 2011;**445**:1625–33.
6. Hill GM, Brewer GJ, Juni JE, Prasad AS. Treatment of Wilson's disease with zinc. II. Validation of oral 64 copper with copper balance. *Am J Med Sci* 1986;**292**:344–9.
7. Ceko MJ, Aitken JB, Harris HH. Speciation of copper in a range of food types by x-ray absorption spectroscopy. *Food Chem* 2014;**164**:50–4.
8. Ohink H, Thiele DJ. How copper traverses cellular membranes through the copper transporter 1, Ctrl. *Ann NY Acad Sci* 2014;**1314**:32–41.
9. Foley PT. International copper demand patterns – the case of plumbing tube. New York (NY): CRU Consultants Inc. Economics of Internationally Traded Minerals, Economics of Copper, Section 5.2. p. 183–186.
10. Ueda K, Kawano H, Hasuo Y, Fujishima M. Prevalence and etiology of dementia in a Japanese community. *Stroke* 1992;**23**:798–803.
11. White L, Petrovitch H, Ros GW, et al. Prevalence of dementia in older Japanese-American men in Hawaii: the Honolulu-Asia Aging Study. *J Am Med Assoc* 1996;**276**:955–60.
12. Brewer GJ. Copper excess, zinc deficiency, and cognition loss in Alzheimer's disease. *Biofactors* 2012;**38**:107–13.
13. Shen XL, Yu JH, Zhang DF, Xie JX, Jiang H. Positive relationship between mortality from Alzheimer's disease and soil metal concentrations in mainland China. *J Alzheimers Dis* 2014;**42**:893–900.
14. Wu M, Han F, Gong W, Feng L, Han J. The effect of copper from water and food: changes of serum nonceruloplasmin copper and brain's amyloid-beta in mice. *Food Funct* 2016;**7**:3740–4.

CHAPTER NINE

Increased Copper Absorption Resulting From Dietary Changes in Developed Countries as Another Causal Factor in the Alzheimer's Disease Epidemic

In the previous chapter the focus has been on the specific cognitive neurotoxicity of copper-2 as causal, at least in part, of the Alzheimer's disease (AD) epidemic. But there is evidence that an increased copper load, deriving in part from increased absorption of copper, irrespective of copper valence, due to dietary changes in developed countries in the 20th century, can also be causal of the AD epidemic. These data are strongly supportive of a key role for copper in general, irrespective of valence, in the pathogenesis of AD. The chapter will begin with data showing AD is a disease of general copper toxicity, and end with data showing that increased body loading of copper is a risk factor for AD, and a discussion of dietary changes that have increased the body loading of copper in the 20th century in developed countries.

DATA THAT SHOW THAT AD IS A DISEASE OF GENERAL COPPER TOXICITY

Data From the Squitti Group Showing AD Pathogenesis is Closely Tied to Elevations in Size of the Blood Free Copper Pool

The Squitti group has made four important points in their publications, as follows:

1. The free copper pool is significantly increased in AD patients.

   What the blood free copper pool is has been explained earlier in the book, but to review briefly, blood copper can be thought of as in two pools. About 85%–90% of blood copper is safely bound in a molecule called ceruloplasmin. The other 10%–15% is loosely bound to various molecules and is called free copper. It is this free copper that

Environmental Causes and Prevention Measures for Alzheimer's Disease
ISBN 978-0-12-811162-8
http://dx.doi.org/10.1016/B978-0-12-811162-8.00009-3

is immediately available for the body's needs. But if this free copper pool becomes expanded, it is this free copper that causes toxicity and damage.

First, it has been shown by the Squitti group[1] that the blood free copper pool is, on average, increased in size in AD patients, in comparison to age-matched controls. This is consistent with copper toxicity occurring in AD.

2. The size of the free copper pool correlates negatively with cognitive measures in AD patients.

Second, the Squitti group[2] has shown that the size of the blood free copper pool correlates negatively with measures of cognition, such as a test called the Standardized Mini–Mental State Examination, or the MMSE, in AD patients. That is, the higher the blood free copper level, the lower the cognition score. This suggests that the higher the level of free copper in the blood, the more damaging it is to cognition.

3. The size of the free copper pool correlates positively with the rate of cognition loss in AD patients.

Third, the Squitti group[3] has shown that the level of blood free copper correlates positively with the rate of cognition loss over time in AD patients. That is, the higher the level of blood free copper, the greater the rate of cognition loss over time. Again, this suggests that the higher the level of free copper, it is more damaging to cognition over time.

4. The size of the free copper pool correlates positively with the probability of conversion of mild cognitively impaired (MCI) patients to full AD.

Fourth, the Squitti group[4] has shown that the higher the blood free copper, the greater the risk that MCI patients (the precursor state to AD where there is only moderate memory loss) will convert to full-blown AD status.

These last three pieces of data show that blood free copper is intimately involved with the pathogenesis and progression of AD. These data do not speak directly to the involvement of inorganic copper, that is, copper-2, since they simply reflect total blood free copper, with no differentiation or measurement of the two types of copper. But they clearly show that AD, at least in a large proportion of patients, is a disease of copper toxicity.

Data Showing Copper Toxicity in AD Brains

James et al.[5] have done some interesting studies where they find an increase in what they called "labile copper" in the brains of AD patients causing toxicity. This is direct data showing increased copper toxicity in the brains of AD patients and further incriminates copper as playing an important role in the pathogenesis of AD. Again it does not discriminate between copper-2 and copper-1. However, again, this data shows that copper toxicity is a major factor in AD pathogenesis.

Summarizing all the data presented so far in this chapter, they provide overwhelming support of the concept that copper toxicity is a major causal factor in AD. Thus, AD joins Wilson's disease (WD) as a disease of copper toxicity.[6] Yet, why are the two diseases so different clinically? WD affecting the brain involves those areas controlling muscle movement, so that coordination, speech, swallowing etc., are affected, but there is little if any effect on cognition. AD affects only cognition and does not affect muscle movement control. The answers may be in the size or intensity of the copper challenge and the duration of the exposure to high copper levels. In Wilson's disease, there is a huge copper challenge, and it appears the movement control parts of the brain are the most vulnerable to such a high and acute level of copper toxicity. The effects of copper on cognition may take a long duration. In Wilson's disease, patients either die young or are treated young, removing the excess copper load. In AD, aging, which is a major risk factor, may be required to work with a more mildly increased copper load over a lifetime to produce AD.

DATA THAT SHOW INCREASED BODY COPPER LOADING IS A RISK FACTOR FOR AD

Squitti et al.[7] have shown that there is an increased prevalence of ATP7B mutant alleles in the AD population. ATP7B is the Wilson's disease gene, and when there are disabling mutations of both copies of this gene, in other words the homozygous state, the serious disease of copper toxicity, Wilson's disease, develops. Heterozygous carriers of one copy of a disabled ATP7B gene have a mild increase in body copper loading, manifested as increased 24h urine copper and increased liver copper content but not enough increased copper to require the type of anticopper treatment carried out for Wilson's disease.

The increased prevalence of ATP7B mutant alleles in the AD population indicates that these alleles confer an increased risk of AD. Assuming that these alleles cause a mildly increased copper load as just described for Wilson's disease causing alleles in the heterozygous state, this indicates a mild increase in body copper loading is a risk for AD. This conclusion has implications of risk for people beyond those who carry an ATP7B allele and would also apply to anyone who has an increase body loading of copper for any reason, including dietary ones. Then the question can be asked, are there any dietary changes in developed countries in the 20th century that could lead to an increased body loading of copper, and thus contribute to the AD epidemic? The answer is probably yes, increased meat eating will do this, because copper is so much better absorbed from meat than from vegetable foods.[8] It is estimated that about 50% more copper is absorbed from a meat-containing diet than from a strictly vegetarian diet. In the next section data indicating a change in the dietary consumption of meat in the 20th century will be considered.

Dietary Changes in Developed Countries That Have Increased Body Copper Loading in the 20th Century

One dietary change in the 20th century is the increased red meat consumption as a result of a huge increase in feed lot usage, primarily for cattle. In the United States for example, by 1950, feed lots with as many as 100,000 cattle were in use.[9] Cattle feeding technology had developed to the point where a 545-pound steer could be brought to slaughter in 14 months, with "well-marbled" (quite fatty) meat. This resulted in 99% of beef consumed in the United States coming from feed lot cattle.[9] This meat also has a much higher fat content than meat previously consumed. Data indicate that the total average meat intake per person in the United States in the 20th century increased by about 50 pounds, an increase of about 35%, compared to the 19th century.[10]

In summary of this area, the increased risk of AD from having an ATP7B mutant allele would not be expected to contribute very much to the AD epidemic, because these alleles are not common enough to contribute substantially. Rather they signal that mildly increased body copper load for a lifetime can increase risk of AD. In turn, this indicates dietary changes in the 20th century, in developed countries, if they increase body copper load, could contribute to the AD epidemic. Since increased meat intake would increase body copper load, and since meat eating has substantially increased in the 20th century, it is concluded that the dietary change of increased meat eating in the 20th century in developed countries has contributed to the AD epidemic.

REFERENCES

1. Squitti R, Pasqualetti P, Dal Forno G, et al. Excess of serum copper not related to ceruloplasmin in Alzheimer's disease. *Neurology* 2005;**64**:1040–6.
2. Squitti R, Barbati G, Rossi L, et al. Excess of nonceruloplasmin serum copper in AD correlates with MMSE, CSF [beta]-amyloid, and h-tau. *Neurology* 2006;**67**:76–82.
3. Squitti R, Bressi F, Pasqualetti P, et al. Longitudinal prognostic value of serum "free" copper in patients with Alzheimer disease. *Neurology* 2009;**72**:50–5.
4. Squitti R, Ghidoni R, Siotto M, et al. Value of serum nonceruloplasmin copper for prediction of mild cognitive impairment conversion to Alzheimer disease. *Ann Neurol* 2014;**75**:574–80.
5. James SA, Voritakis I, Adlard PA, et al. Elevated labile Cu is associated with oxidative pathology in Alzheimer disease. *Free Radic Biol Med* 2012;**52**:298–302.
6. Brewer GJ. Wilson's disease. In: Kasper DL, Fauci AS, Hauser SL, Longo DL, Jameson JL, Loscalzo J, editors. *Harrison's principles of internal medicine*. 19th ed. New York: McGraw-Hill Companies, Inc.; 2015.
7. Squitti R, Polimanti R, Siotto M, et al. ATP7B variants as modulators of copper dyshomeostasis in Alzheimer's disease. *Neuromol Med* 2013;**15**:515–22.
8. Brewer GJ, Yuzbasiyan-Gurkan V, Dick R, Wang Y, Johnson V. Does a vegetarian diet control Wilson's disease? *J Am Coll Nutr* 1993;**12**:527–30.
9. Cordain L, Eaton SB, Sebastian A, et al. Origins and evolution of the Western diet: health implications for the 21st century. *Am J Clin Nutr* 2005;**81**:341–54.
10. Teicholz N. How Americans got red meat wrong. *Atlantic* June 2, 2014.

CHAPTER TEN

The Copper Hypothesis Fits Nicely With Known Risk Factors and Theories of Alzheimer's Disease Causation

INTRODUCTION

So far in this book the case has been made that AD is in large part a copper toxicity disease, at least in developed countries. It has been shown that copper toxicity comes from copper-2 ingestion and from an increased body copper load for a lifetime. In this chapter, the objective is, first to examine whether noncopper risk factors interact with copper in AD pathogenesis and second whether copper toxicity fits with current theories of primary AD causation.

THE RELATION OF COPPER TOXICITY TO THE IMPORTANT RISK FACTORS FOR AD

Most of the known important risk factors for AD other than copper ingestion were discussed in Chapter 2 and are listed in Table 10.1.

The first risk factor is age. This is a major AD risk factor, but as far as is known, there is no particular relation to copper toxicity, except, of course, the older the person, the more time any toxicity, including copper toxicity, has to act.

The next listed risk factor is possession of an apolipoprotein E4 allele.[1,2] There are three apolipoprotein alleles in most populations, E2, E3, and E4. In the United States, 2% of the population is homozygous for E4, and their risk of AD is increased 8- to 12-fold. About 23% of the population carries one copy of E4, and their risk of AD is increased threefold.[3] E3 seems to be neutral with respect to risk for AD, and E2 seems to be somewhat protective against risk for AD. Somewhere between 40% and 65% of AD patients carry at least one E4 allele.[3] Thus, it is a major risk for AD.

Environmental Causes and Prevention Measures for Alzheimer's Disease
ISBN 978-0-12-811162-8
http://dx.doi.org/10.1016/B978-0-12-811162-8.00010-X

Table 10.1 The More Important Risk Factors for AD

1	Age
2	Possession of an apolipoprotein E4 allele
3	Possession of an ATP7B mutant allele
4	Possession of a hemochromatosis or transferrin mutant allele
5	Positive family history of AD

With respect to interaction with copper, the E2 allele produces a protein that contains two cysteines that can bind copper, the E3 allele produces a protein with one copper-binding cysteine, and the E4 allele produces a protein with no copper-binding cysteines. It is believed the apolipoproteins are involved in copper transport, including copper removal from the brain, and obviously the E2-produced protein would be best at this, the E3 protein next best, and the E4 protein the poorest. Thus, it is clear that there is an important interaction between copper and a major risk factor for AD, the apolipoprotein E4 allele. This helps strengthen the rationale for the copper hypothesis in AD causation.

The next risk factor listed in Table 10.1 is possession of an ATP7B mutant allele. The Squitti group[4,5] has published extensively on the increased prevalence of ATP7B mutant alleles in the AD patient population. This increased prevalence means the ATP7B alleles increase AD risk.

The connection of increased risk from possession of an ATP7B allele with the copper hypothesis is obvious. ATP7B is, of course, the Wilson's disease gene. Wilson's disease is an inherited disease of copper accumulation and extreme copper toxicity.[6] It is a recessive disease, meaning that both copies of ATP7B have to be disabled to cause the disease. It is known that carriers of one copy of a Wilson's disease causing mutation in ATP7B have a mildly increased body copper load, with significantly increased levels of liver and urine copper. It seems likely that the ATP7B alleles found in increased prevalence in AD populations also affect the function of the gene, causing a mildly increased body copper load and increasing risk of AD. Again this fits nicely with the copper hypothesis of AD causation.

A question that arises here is why Wilson's disease, with its extremely high free copper levels and severe copper toxicity, does not cause AD. Instead the brain manifestations of copper toxicity in Wilson's disease are a movement disorder, with no effect on cognition. The likely explanation is that patients with Wilson's disease die young or are treated at a young age to remove the copper burden. The parts of the brain that are susceptible to an acute high level of copper, intense but relatively short-lived, are those that control movement, not cognition. Thus, age, a major AD risk factor, does not play a role here. In contrast, in carriers of an ATP7B mutant allele, there is a lifetime of modestly increased body copper load, allowing age, a major risk factor for AD, to interact with elevated copper and less intense copper toxicity to produce AD.

The next risk factor in Table 10.1 is possession of a hemochromatosis[7] or transferrin[8] mutant allele. Both of these genes influence iron metabolism and iron levels. Since excessive levels of iron cause toxicity by oxidant damage, and since copper causes toxicity by oxidant damage as well, the toxicity of iron fits with the copper toxicity hypothesis in that it is a second way in which a metal produces oxidant stress.

The next risk factor in Table 10.1 is a positive family history for AD.[3] If one first degree relative (sibling or parent) has AD, risk of AD is increased, and if two first degree relatives have AD, the risk is even higher. This risk comes from a combination of genetic and environmental factors. The apolipoprotein E4 allele risks are part of the genetic risk, and there are numerous other genetic areas that have been associated with risk of AD in genome wide association studies (GWAS). Each of these areas individually adds a tiny amount of risk, too small to consider individually. However, it is the environmental area that needs to be considered here. These first degree relatives would all share drinking water, with its associated copper-2 content. They would also likely share tendencies to ingest multimineral supplement pills containing copper-2. In this way, the increased risk of AD from a positive family history of AD could be due, in part, to shared increased levels of copper-2 ingestion.

THE RELATION OF A CAUSAL ROLE FOR COPPER TO THE CURRENT THEORIES OF PRIMARY CAUSATION OF AD

The Amyloid Cascade Hypothesis of Primary AD Causation

The most widely accepted theory of primary AD causation is the amyloid cascade hypothesis.[9] The basics of this theory were portrayed in Fig. 2.1. In this theory, an enzyme called beta secretase clips off a piece of the protein called amyloid precursor protein (APP). The piece clipped off is called beta amyloid (Aβ). This process goes on in the normal brain, and Aβ is cleared as it is formed. The normal function of Aβ is unknown. In the AD brain, Aβ accumulates for unknown reasons, possibly because it is formed faster or cleared slower. The accumulating Aβ aggregates, and the aggregates are called amyloid plaques. The plaques can bind iron and copper and this causes emission of damaging oxidant radicals. It is known that oxidant damage is occurring in the AD brain, and this could be at least one source.

The primary causative role of Aβ in AD causation is supported by the consistent presence of amyloid plaques in the brains of AD patients at autopsy. It is also supported by the occurrence of early onset AD when there are certain mutations in the APP gene.[10] This is called familial AD and accounts for about 1% of AD patients. The fact that mutations in the APP gene can cause AD lends support to a primary role for the amyloid cascade hypothesis.

Copper has two potential roles in the amyloid cascade hypothesis. The first is that copper can cause aggregation of Aβ into plaques.[11] The second is that once plaques are formed, copper can bind to the plaques and cause emission of damaging oxidant radicals.[12]

To be clear, under the amyloid cascade hypothesis, the copper hypothesis does not claim that copper is a primary cause of AD. Accumulation and aggregation of Aβ is the primary cause. But copper serves as a trigger of the disease, causing Aβ aggregation and/or greatly increased emission of damaging oxidant radicals. Through this triggering mechanism, it is proposed by the copper hypothesis that copper is a major cause of the AD epidemic in developed countries over the last century.

The Oxidant Damage Hypothesis of Primary AD Causation

Not all accept the amyloid cascade hypothesis. An attractive alternative hypothesis, championed by Perry and his group, called here the primary oxidant damage hypothesis, has been proposed. Under this hypothesis, the primary event is oxidant damage.[13] It is known that the AD brain exhibits oxidant damage. Under this hypothesis it is suggested that Aβ plays a protective role, acting as an antioxidant. Aβ can also reduce toxic copper-2 to the safer copper-1. Under this hypothesis, as oxidant stress continues, more Aβ is produced to protect the brain. As the Aβ accumulates, it aggregates and forms amyloid plaques. Under this hypothesis, the amyloid plaques are the proverbial cart, not the horse as proposed in the amyloid cascade hypothesis.[14] In other words, the plaques are the result of damage, not the cause.

Under the primary oxidant damage hypothesis, copper could be the major and primary cause of the oxidant damage. So, in this case, copper could be a primary AD causative agent and cause the AD epidemic in developed countries over the last century in that manner.

SUMMARY

In this chapter, first it has been shown how the copper hypothesis may partially explain some of the risk factors for AD. This fit helps support the copper hypothesis. Second, how the copper hypothesis fits with the two most accepted theories of primary AD causation has been discussed. In the amyloid cascade hypothesis, copper could play an important role in triggering the cascade. In the primary oxidant damage hypothesis, copper could be the cause of the primary oxidant damage.

REFERENCES

1. Sauners AM, Strittmatter WJ, Schmechel D, et al. Association of apolipoprotein E allele epsilon 4 with late-onset familial and sporadic Alzheimer's disease. *Neurology* 1993;**43**:1467–72.
2. Farrer LA, Cupples LA, Haines JL, et al. Effects of age, sex, and ethnicity on the association between apolipoprotein E genotype and Alzheimer disease: a meta–analysis. *JAMA* 1997;**278**:1349–56.
3. Alzheimer's Disease Association. *Alzheimer's disease facts and figures*. 2016.
4. Squitti R, Polimanti R, Bucossi S, et al. Linkage disequilibrium and haplotype analysis of the ATP7B gene in Alzheimer's disease. *Rejuvenation Res* 2013;**16**:3–10.
5. Squitti R, Polimanti R, Siotto M, et al. ATP7B variants as modulators of copper dyshomeostasis in Alzheimer's disease. *Neuromol Med* 2013;**15**:515–22.
6. Brewer GJ. Wilson's disease. In Harrison's principles of internal medicine. In: Kasper DL, Fauci AS, Hauser SL, Longo DL, Jameson JL, Loscalzo J, editors. 19th ed. New York: McGraw-Hill Companies, Inc.; 2015.
7. Moalem S, Percy ME, Andrews DF, et al. Are hereditary hemochromatosis mutations involved in Alzheimer disease? *Am J Med Genet* 2000;**93**:58–66.
8. Zambenedetti P, de Bellis G, Biunno I, Musicco M, Zatta P. Transferrin C2 variant does confer a risk for Alzheimer's disease in Caucasians. *J Alzheimers Dis* 2003;**5**:423–7.
9. Hardy J, Selkoe DJ. The amyloid hypothesis of Alzheimer's disease: progress and problems on the road to therapeutics. *Science* 2002;**297**:353–6.
10. Chartier-Harlin MC, Crawford F, Houlden H, et al. Early-onset Alzheimer's disease caused by mutations at codon 717 of the β-amyloid precursor protein gene. *Nature* 1991;**353**:844–6.
11. Sarell CJ, Wilsinson SR, Viles JH. Substoichiometric levels of Cu^{2+} ions accelerate the kinetics of fiber formation and promote cell toxicity of amyloid- β from Alzheimer disease. *J Biol Chem* 2010;**285**:41533–40.
12. Sayre LM, Perry G, Harris PL, Liu Y, Schubert KA, Smith MA. In situ oxidative catalysis by neurofibrillary tangles and senile plaques in Alzheimer's disease: a central role for bound transition metals. *J Neurochem* 2000;**74**:270–9.
13. Nunomura A, Tamaoki T, Motohashi N, et al. The earliest stage of cognitive impairment in transition from normal aging to Alzheimer disease is marked by prominent RNA oxidation in vulnerable neurons. *J Neuropathol Exp Neurol* 2012;**71**:233–41.
14. Lee HG, Castellani RJ, Zhu X, Perry G, Smith MA. Amyloid- β in Alzheimer's disease: the horse or the cart? Pathogenic or protective? *Int J Exp Pathol* 2005;**86**:133–8.

CHAPTER ELEVEN

Prevention Measures Action Items: Two Simple Steps to Eliminate Ingestion of Copper-2, and Dietary Changes to Reduce Copper Absorption

INTRODUCTION

In this book so far, strong efforts have been made to show that developed countries are in the midst of a severe AD epidemic which has blossomed in the last 100 years and is almost certainly due to some change in the environment. Equally strong effort has been made to show almost overwhelming evidence that the environmental culprit was ingestion of copper, primarily copper-2 but also increased absorption of copper in general because of dietary changes. Earlier (Chapter 2) it was illustrated what a disastrous disease AD is, robbing a high percentage of seniors of their "golden years," their retirement years, earned after a lifetime of work. Their memories even of loved ones, their ability to carry out even simple tasks, all their enjoyments, stolen from them. The family is robbed of the companionship and love from the patient, and must shoulder the burden of 24/7 care, and then the financial burden of the patient's institutionalization. AD is indeed a terrible disease, robbing people of their humanity. They can be compared to a pet animal. And now AD has become so common.

Although the evidence for the copper-2 hypothesis is very strong, it is not well enough known, or finally proven to the satisfaction of all experts, so there is no ground swell among physicians, public health officials, and the various branches of government to eliminate or severely limit ingestion of copper-2. So it is up to the individuals to take action to protect themselves if they are convinced by the strong evidence indicting copper-2 for AD causation put forth in this book. The situation can be compared to the situation when it was first proposed that cigarette smoking caused lung cancer. At that time this hypothesis had accumulated strong evidence, but was

Environmental Causes and Prevention Measures for Alzheimer's Disease
ISBN 978-0-12-811162-8
http://dx.doi.org/10.1016/B978-0-12-811162-8.00011-1

not finally proven. Those who stopped smoking at that point gained great benefit. Now, even smokers are aware of the established fact that cigarette smoking is an important cause of lung cancer and heart disease. The point is, it seems very likely that copper-2 ingestion will be established as a major cause of AD, and those who avoid it now have an excellent chance to be greatly benefitted.

So in the next section, the two rather simple steps to eliminate or greatly reduce copper-2 ingestion will be presented. Since it appears that an increased body copper load in general, irrespective of valence, is also a risk factor for AD, the section after that will discuss dietary changes to reduce overall copper body loading.

TWO SIMPLE STEPS TO ELIMINATE INGESTION OF COPPER-2

The two major sources of copper-2 ingestion are from copper-containing supplement pills and from drinking water. How to eliminate ingestion of copper-2 from these sources will now be discussed.

Almost all multimineral and multivitamin/multimineral supplement pills contain copper-2. If there is any question about this for a given product, consult the label. The label will just say copper; it will not say copper-2, but it is all copper-2 nonetheless. If any of these copper-2-containing products are in the home, they should be thrown out, because they are a form of poison. In its own way, the copper in these pills is just as toxic as lead, and no one would knowingly ingest a lead-containing pill. Similarly, copper-containing supplement pills should not be purchased. Multivitamin preparations are safe. If one needs a mineral, such as calcium, iron, zinc, etc., it should be purchased as a separate item, that is, a pill containing just that mineral. Despite the claims of some, the general population does not need copper supplementation. Copper deficiency is very rare. For the occasional person who is copper deficient, and these are well-identified patients who have had bowel surgery or have malabsorption problems, there is a company producing copper-1-containing pills. So even these patients no longer have to take the toxic copper-2 pills.

Bottom line, the first step in eliminating copper-2 ingestion is very simple. Just avoid taking copper-2-containing supplement pills.

The second step, minimizing copper-2 ingestion from drinking water is a little harder but still relatively straightforward. According to studies in North America, the drinking water copper levels of about two-thirds of households are unsafe,[1] if the AD animal models where copper levels at 0.12 ppm were found to be unsafe are a good guide.[2-4] In these studies, drinking water in about one-third of households contained copper levels of over 0.1 ppm, found unsafe in the animal models, about one-third were under 0.01 ppm and deemed safe, and about one-third were between these levels and of unknown safety because they have not been tested.[1] So, to be on the safe side, it is recommended here that drinking water with copper levels above 0.01 ppm be viewed as unsafe. That means that about two-thirds of North American households have drinking water with unsafe levels of copper.

So obviously, the first thing that must be done is to test the drinking water copper levels. (To be completely clear, all copper in drinking water is copper-2.) The copper levels should be tested, even if the dwelling does not have copper plumbing, because the source water may have elevated copper levels. There are many companies that can be found online that offer assay of substances in water, including copper. If levels are 0.01 ppm (0.1 µg/L) or less, they are safe. If over 0.01 ppm, the copper plumbing (if present) need not be removed. A copper-removing device, such as a reverse osmosis device, can be placed on the tap used for drinking and cooking water. This will reduce the copper-2 to safe levels. Installing such a device is not very expensive.

Thus, by following these two simple steps, both of which can be carried out without too much difficulty or expense, the ingestion of copper-2 can be greatly reduced. It cannot be eliminated completely, because there remains a little copper-2 in drinking water and a little in food. But reduction in copper-2 ingestion back to 19th century levels should eliminate much of the great increase in AD prevalence of the AD epidemic.

DIETARY CHANGES TO REDUCE COPPER ABSORPTION

To briefly review, Squitti et al.[5,6] have found an increased prevalence of ATP7B alleles in AD populations. ATP7B is the Wilson's disease gene. Wilson's disease is an autosomal recessive disease of copper toxicity.[7] Carriers of one disabled copy of an ATP7B allele have a mild increase in body copper loading manifested by an increased liver and urine copper. Assuming the ATP7B alleles found at increased prevalence in AD populations cause a mild increase in body copper loading, this indicates an increase in body copper loading from any source, such as dietary sources, would also be an AD risk factor. Looking at dietary changes in the 20th century which might increase body copper loading and thus contribute to the AD epidemic, increased meat eating is an obvious dietary change which could increase body copper loading, as discussed in Chapters 5 and 9.

The mechanism of increased body copper loading from increased meat eating is that copper is so much better absorbed from meat than from vegetable foods.[8] Estimates are that there is a 50% increase in copper absorption from a meat-containing diet compared to a vegetarian diet. The main reasons for increased meat eating in developed countries in the 20th century is the major development of feed lot technology[9] as described in Chapter 9 and the increase in poultry consumption. It is estimated that there has been a resulting 35% increase in meat eating in the 20th century compared to the 19th century.[10]

From this, it is clear that the dietary change that is necessary to reduce copper absorption and thus body copper loading is to reduce meat eating. The best guess on how much to reduce meat eating might be to average a 35% reduction, since that is the estimated average increase in meat ingestion in the 20th century.[10] This could be accomplished by making 35% of meals, particularly the bigger meals such as dinner, meatless, or by reducing all meat portions by 35%. From the overall health standpoint, a reduction in meat eating is a healthy step. It has been shown that a 50% reduction in meat eating reduces overall mortality by 42%.[11]

It is realized that this change in dietary habits is more difficult than the steps to reduce copper-2 ingestion, because it is a lifestyle change, and these are always more difficult. A possible alternative is to take a zinc supplement. Taking a zinc pill of up to about 50 mg/day (away from food by 2 h) is probably effective in lowering free copper a little and is safe for most people in terms of not producing clinical copper deficiency. In view of the relative difficulty of lowering free copper levels by either lowering meat eating or taking a zinc supplement, if we assume that the copper-2 hypothesis together with the more general AD causation by copper hypothesis are responsible for the AD epidemic, the question of the size of a positive effect from dietary change to reduce copper absorption, or taking a zinc pill, depends on the quantitative effect of each of the two causes. At this time it is not possible to determine the quantitative importance of copper-2 reduction versus reduction in total body copper loading via a reduction in meat eating or taking zinc. Given the importance of prevention of the disease, the safest thing at this point is to do both.

Another important question is, what is the overall expectation if both of these steps are accomplished? The best expectation would be to reduce the high AD prevalence of the AD epidemic. It would be expected that there would still be a prevalence of, say 1%, in those over age 65, as there was in the 19th century, resulting from risk factors unrelated to copper. If that occurred it means that only about 5% of those now slated to get AD would get the disease, a 20-fold reduction in risk.

REFERENCES

1. Brewer GJ. Copper excess, zinc deficiency, and cognition loss in Alzheimer's disease. *Biofactors* 2012;**38**(2):107–13.
2. Sparks DL, Schreurs BG. Trace amounts of copper in water induce beta-amyloid plaques and learning deficits in a rabbit model of Alzheimer's disease. *Proc Natl Acad Sci USA* 2003;**100**:11065–9.
3. Sparks DL, Friedland R, Petanceska S, et al. Trace copper levels in drinking water, but not zinc or aluminum, influence CNS Alzheimer-like pathology. *J Nutr Health Aging* 2006;**10**:247–54.
4. Singh I, Sagare AP, Coma M, Perlmutter D. Low levels of copper disrupt brain amyloid-beta homeostasis by altering its production and clearance. *Proc Natl Acad Sci USA* 2013;**110**:14471–6.
5. Squitti R, Polimanti R, Bucossi S, et al. Linkage disequilibrium and haplotype analysis of the ATP7B gene in Alzheimer's disease. *Rejuvenation Res* 2013;**16**:3–10.
6. Squitti R, Polimanti R, Siotto M, et al. ATP7B variants as modulators of copper dyshomeostasis in Alzheimer's disease. *Neuromol Med* 2013;**15**:515–22.
7. Brewer GJ. Wilson's disease. In: Kasper DL, Fauci AS, Hauser SL, Longo DL, Jameson JL, Loscalzo J, editors. *Harrison's principles of internal medicine*. 19th ed. New York: McGraw-Hill Companies, Inc.; 2015.
8. Brewer GJ, Yuzbasiyan-Gurkan V, Dick R, Wang Y, Johnson V. Does a vegetarian diet control Wilson's disease? *J Am Coll Nutr* 1993;**12**:527–30.
9. Cordain L, Eaton SB, Sebastian A, et al. Origins and evolution of the Western diet: health implications for the 21st century. *Am J Clin Nutr* 2005;**81**(2):341–54.
10. Teicholz N. How Americans got red meat wrong. *Atlantic* June 2, 2014.
11. Sinha R, Gross AJ, Graubard BI, Leitzmann MF, Schotzkin A. Meat intake and mortality: a prospective study of over half a million people. *Arch Int Med* 2009;**167**:562–71.

CHAPTER TWELVE

Failures: What the Government Has Not Done to Ensure Healthy Drinking Water and Nontoxic Multimineral Pills

INTRODUCTION

Chapter 11 went into considerable detail about what the individual can do to protect themselves against ingestion of toxic copper-2 in drinking water and toxic copper-2 in supplement pills. But it is likely that a large segment of the population either will not get this message or if they do will not be motivated to act on it. So, the bottom line is, much of the population will probably remain at risk of having Alzheimer's disease (AD) as a result of ingesting toxic copper-2. So a legitimate question can be raised, does not the government have a responsibility to protect its citizens from poisons like copper-2? The answer is clearly yes. However, there has been no government action. So the next question is, why has there been no government action?

To answer that question, one has to examine what it takes to get government action. If scientists find that an agent causes cancer, and then it is found that that substance is getting into a water supply, government acts quickly to eliminate that substance from the water supply or to severely limit allowable levels. Similarly, if a type of pill, whether made by a pharmaceutical company or an over-the-counter supplier is found to be contaminated with something toxic, the government moves quickly to make sure contaminated pills are pulled off the shelves and that the problem at the company is solved.

All would agree that AD is as serious a disease as cancer or any disease that would be caused by contaminants in pills. So the problem is not the severity or the frequency of the disease being caused. That means the problem lies in the appropriate government agencies not being aware that copper-2 is causing the disease. Or if they are aware of the hypothesis that copper-2 is causing AD, they have not yet accepted it. There are several reasons they might not have accepted it. First, they may not be aware how strong the evidence is. This is probably true of most of them. Second, they

Environmental Causes and Prevention Measures for Alzheimer's Disease
ISBN 978-0-12-811162-8
http://dx.doi.org/10.1016/B978-0-12-811162-8.00012-3

know that there have been other claims of environmental causation of AD; aluminum is a good example and simply believe copper is just this year's favorite metal. Third, there may be pressure such that they do not want to believe the hypothesis is true. For example, not allowing copper-2 in supplement pills would force multimineral pill makers to spend a lot of money reformulating and possibly rebranding their product. Many drinking water supply systems across the country would have to work hard and spend money, to get copper levels down to a new required low level, such as down to 0.01 ppm from the currently allowed 1.3 ppm. And fourth, they may just plain doubt that the scientific evidence is strong enough to prove the case.

The fourth reason is the only legitimate one. But there will always remain doubters of the validity of a given hypothesis, and the government must not wait to satisfy all doubters that a problem exists. The first step by scientists and the educated community should be to eliminate the first three reasons for not acting. This can be done by more extensive education of the decision makers, and by public pressure.

This book is designed to educate academicians who can help educate the public and decision makers. This chapter is designed to inform all of what needs to be done at the government level.

GOVERNMENT FAILURE NUMBER ONE: THE FOOD AND DRUG ADMINISTRATION (FDA) HAS FAILED TO ELIMINATE COPPER-2 FROM SUPPLEMENT PILLS, OR AT LEAST STUDY THE ISSUE

Almost all multimineral and multivitamin/multimineral supplement pills contain copper, usually in a dose of about 1 mg/pill. These are over-the-counter products, widely advertised, and widely taken by adults, motivated by advertising that people need to take such pills to make sure they have no deficiencies. The public believes these pills are completely safe and, further, that the pills will assure them of continued health by avoidance of deficiencies.

So the public has put their trust in these pills because of the advertising and because there is a widespread belief and trust that the government is watching and will make sure that the public is protected from unsafe products. But are multimineral pills-containing copper safe? The 2006 paper by Morris et al.[1] provided strong evidence that ingestion of copper-containing pills is not safe. They cause rapid cognition loss, and likely AD. These data have been added to by the study of Mursa et al.,[2] in which they found ingestion of copper pills in women in Iowa was associated with a 42% increase in mortality.

Both the Morris et al.[1] and Mursa et al.[2] studies showing an association between ingestion of copper pills and bad things happening does not prove the copper pills caused the bad things. Association does not prove causality. Perhaps this is why the FDA has ignored these studies, if in fact, the studies have even come to their attention. But there have been two studies (Morris et al.[1] and Mursa et al.[2]) plus the numerous publications the author has put together,[3–8] building a strong case for the toxicity of copper-2 in these supplement pills as causing serious and frequent disease, and the FDA should at least examine the issue. A minimal step would be to recommend and help organize a replication of the Morris et al.[1] study with appropriate placebo controls. A more aggressive, but probably appropriate, step would be to ban copper-2 in supplement pills unless a given manufacturer could prove that it is safe, i.e., did not cause cognition loss or excess mortality. Copper-1-containing pills are now being made and could be used for those relatively rare patients who actually need copper supplementation.

The recommendation made here is as follows:

It is recommended that the FDA appoint a commission with good representation of knowledgeable scientists to study the issue of the safety of copper-2 in supplement pills and make recommendations about what further actions should be taken by the FDA. It is further requested that the FDA comply with the commission recommendations.

GOVERNMENT FAILURE NUMBER TWO: THE ENVIRONMENTAL PROTECTION AGENCY HAS FAILED TO LIMIT COPPER IN DRINKING WATER TO A SAFE LEVEL TO ENSURE HEALTHY DRINKING WATER

The pioneering studies of Sparks and Schreurs[9] show how dramatically toxic copper-2 in drinking water is. At 0.12 ppm concentrations in the drinking water of an AD animal model, copper-2 greatly enhanced amyloid plaque formation and memory loss in the animals. And this study is far more than an association study. It is a randomly controlled trial (an RCT) in animals. One group of animals had distilled water for drinking and another group had the same distilled water with 0.12 ppm copper. In subsequent studies, Sparks et al.[10] showed that only copper, not aluminum or zinc, would cause the strong development of AD in the animals. This is a true RCT, the highest level of evidence in humans. Only the copper, not placebo (distilled water) or aluminum and zinc as further controls, caused the disease. Sparks et al.[10] also showed that the 0.12 ppm copper in the drinking water would enhance AD in several animal species, including mice and dogs, as well as rabbits. This work was confirmed in a mouse model in a second laboratory.[11]

The work described above comes as close as one can get in animals to proving that trace amounts of copper-2 in drinking water causes AD in humans. The only thing better would be to repeat the experiment in humans, which would be unethical. But in fact, humans have done something similar to themselves, as shown in the studies of Morris et al.[1] in which copper-2 ingestion in supplement pills was associated with, and probably caused, marked cognition loss.

The Environmental Protection Agency (EPA) of the US government is responsible for setting allowable limits for copper in drinking water. And for many years, the allowable limits for copper have been set at 1.3 ppm,[12] 10 times the amount found toxic and AD causative in animal model studies. So why does not the EPA lower the allowable limits to an "AD safe" level, such as 0.01 ppm? Surely the evidence as said above is clear that drinking water copper levels of 0.1–1.3 ppm are "AD unsafe." Yet for over a decade since that evidence was developed, the EPA has done nothing.

Perhaps the EPA decision makers are unaware of the totality of this evidence, although the author sent the EPA Director in office several years ago a letter and detailed information about the allowable level being much too high because of the AD risk. Perhaps the reasoning is that much of the copper problem in drinking water is coming from the copper plumbing in the home and not from the source water to which the allowable limits apply. But this reasoning is flawed, because the copper levels in the source water in some parts of the country approaches the 1.3 ppm level. Besides, if the EPA set the allowable limit at 0.01 ppm, it would give the citizen a well-publicized "do not exceed" target level for the copper level in drinking water coming out of their tap.

But the most likely explanation for EPA inaction is that since the EPA Director is a political appointment, the EPA responds primarily to political pressure. For example, a few years ago, some senators from Midwestern states, where copper levels in source water are high, asked for allowing a higher level than 1.3 ppm. The major rationale of the request was that it was expensive for water plants in those states to get the source water copper levels down to 1.3 ppm. A committee was set up, with the author one of the members, to examine the safety of increasing the limits.[12] This was before the work of Sparks and Schreurs,[9] so the safety with respect to AD risk was not under consideration. The commission recommended that the limits not be raised, with one of the considerations being increased risk to Wilson's disease patients and carriers of the Wilson's disease gene. Formation of this committee to study the issue raised by the senators is an example of the EPA response to political pressure.

With this last consideration in mind, effort will be made to get exposure of the material in this book to legislators and other decision makers, particularly those who, perhaps for personal reasons, such as a close relative or friend developing AD, have a strong interest in preventing AD. Hopefully, academicians reading this book, will also communicate the message to such people with whom they have contact.

The recommendation here is that the EPA set the allowable limit of copper in drinking water to 0.01 ppm. This is believed to be a very "AD safe" level.

Summarizing, the two recommendations for government action are:

1. It is recommended that the FDA appoint a commission with good representation of knowledgeable scientists to study the issue of the safety of copper-2 in supplement pills and make recommendations about what further actions should be taken by the FDA. It is further requested that the FDA comply with the commission recommendations.
2. It is recommended that the EPA set the allowable limit of copper in drinking water to 0.01 ppm. This is believed to be an "AD safe" level.

REFERENCES

1. Morris MC, Evans DA, Tangney CC, et al. Dietary copper and high saturated and trans fat intakes associated with cognitive decline. *Arch Neurol* 2006;**63**:1085–8.
2. Mursa J, Robein K, Hamack LJ, Park K, Jacobs DR. Dietary supplements and mortality rate in older women: the Iowa Women's Health Study. *Arch Int Med* 2011;**445**:1625–33.
3. Brewer GJ. The risks of copper toxicity contributing to cognitive decline in the aging population and to Alzheimer's disease. *J Am Coll Nutr* 2009;**28**:238–42.
4. Brewer GJ. Issues raised involving the copper hypotheses in the causation of Alzheimer's disease. *Int J Alzheimers Dis* 2011:537528.
5. Brewer GJ. Copper excess, zinc deficiency, and cognition loss in Alzheimer's disease. *Biofactors* 2012;**38**:107–13.
6. Brewer GJ. Copper toxicity in Alzheimer's disease: cognitive loss from ingestion of inorganic copper. *J Trace Elem Med Biol* 2012;**26**:89–92.
7. Brewer GJ. Divalent copper as a triggering agent in Alzheimer's disease. *J Alzheimers Dis* 2015;**46**:593–604.
8. Brewer GJ. Copper-2 ingestion, plus increased meat eating leading to increased copper absorption, are major factors behind the current epidemic of Alzheimer's disease. *Nutrients* 2015;**7**:10053–64.

9. Sparks DL, Schreurs BG. Trace amounts of copper in water induce beta-amyloid plaques and learning deficits in a rabbit model of Alzheimer's disease. *Proc Natl Acad Sci USA* 2003;**100**:11065–9.

10. Sparks DL, Friedland R, Petanceska S, et al. Trace copper levels in the drinking water, but not zinc or aluminum, influence CNS Alzheimer-like pathology. *J Nutr Health Aging* 2006;**10**:247–54.

11. Singh I, Sagare AP, Coma M, et al. Low levels of copper disrupt brain amyloid-beta homeostasis by altering its production and clearance. *Proc Natl Acad Sci* 2013;**110**:14471–6.

12. National Research Council. *Copper in drinking water*. Washington (DC): National Academy Press; 2000.

CHAPTER THIRTEEN

Treatment of Alzheimer's Disease

INTRODUCTION

The main thrust of this book has been in prevention of Alzheimer's disease (AD). However, preventing AD is of little help to those who already have it. The prevention measures that were elaborated on in Chapter 11 are not likely to reverse the disease if it is already well established, although it is possible they might slow its progression.

So what treatment measures are available for those who have already developed AD? Unfortunately, the "treatment cupboard" for AD is almost empty. The situation can be summarized by saying there is no effective treatment for AD, if by "effective treatment" is meant a treatment that will cure the disease, or at least strongly slow its progress.

There are some things that will help a little. In Chapter 5, we discussed a number of lifestyle factors that can either delay initiation of the disease a little, or mitigate it a little once it occurs, or both. These included physical exercise, probably problem-solving mind games, and ingestion of certain nutrients. (These nutrients are all listed in Table 5.1 of Chapter 5.) But none of these will markedly slow the progression of the disease. In the next section we will review some drug treatments which can improve certain symptoms, at least temporarily. But as with the lifestyle factors, none of these can markedly halt the progression of the disease.

What about efforts by the pharmaceutical industry to develop a cure for AD, or at least a drug that would significantly affect disease progression? There has been considerable effort. But unfortunately, the past record is full of the corpses of unsuccessful trials and efforts to develop an effective treatment for AD. These extensive efforts will be reviewed in a later section of this chapter. Despite all the failures so far, there is possibly some hope that an effective treatment may be available sooner rather than later. This will be discussed further near the end of this chapter.

Environmental Causes and Prevention Measures for Alzheimer's Disease
ISBN 978-0-12-811162-8
http://dx.doi.org/10.1016/B978-0-12-811162-8.00013-5

DRUG TREATMENTS WHICH CAN TEMPORARILY IMPROVE CERTAIN SYMPTOMS

The US Food and Drug Administration (FDA) has approved five drugs of this type, all shown in Table 13.1. Three of these drugs are cholinesterase inhibitors, one is an N-methyl D-aspartate (NMDA) antagonist, and one is a combination of these two types.

Cholinesterase is an enzyme that breaks down acetyl choline, an important neurotransmitter in the brain that affects memory and thinking. It appears that the AD brain is not producing adequate acetyl choline, so drugs that inhibit cholinesterase inhibit the breakdown of acetyl choline and make it more available, thus improving some brain functions. The effect is temporary because after a time the AD brain makes so little acetyl choline that inhibiting its breakdown no longer has an effect.

As mentioned, one of the approved drugs, memantine, is an NMDA inhibitor, which works by helping to regulate glutamate activity. Glutamate is another important neurotransmitter, involved in learning and memory. Attachment of glutamate to cell surface "docking sites", which are really NMDA receptors, permits calcium to enter the cell, important in signaling. In AD, excess glutamate can be released, called glutamate excessive cytoactivity, which can damage and even kill neurons. Memantine

Table 13.1 Current US Food and Drug Administration-Approved Drugs for Treating Alzheimer's Disease

Drug Name	Trade Name	Mechanism of Action	Use	Approval Date
Donepezil	Aricept	Cholinesterase inhibitor	Mild, moderate, and severe AD	1996
Rivastigmine	Exelon	Cholinesterase inhibitor	Mild to moderate AD	2000
Galantamine	Razadyne	Cholinesterase inhibitor	Mild to moderate AD	2001
Memantine	Namenda	NMDA antagonist	Moderate to severe AD	2003
Memantine plus Donepezil	Namzaric	NMDA antagonist and cholinesterase inhibitor	Moderate to severe AD	2014

partially blocks the NMDA receptors and helps prevent this damaging chain of events.

The fifth approved drug, trade name Namzaric, combines one of the cholinesterase inhibitors with memantine.

As mentioned, the effect of these drugs is to improve memory and cognition a little, and the effect is temporary. However, this small improvement might be quite significant for the home caregiver. For example, if the patient is able to dress themselves and carry out tasks like tooth brushing and shaving while taking the drug, but not without it, this significantly reduces the amount of time and effort the caregiver has to expend to take care of the patient. This kind of improvement can also delay the requirement for institutionalization, which decreases cost to family and/or insurers. Of course, this is a trade-off, because it lengthens the time the patient remains in the home, thereby lengthening the time the home caregiver has to spend taking care of the patient.

APPROACHES WHICH CAN HELP WITH CERTAIN BEHAVIORS

Many caregivers say that the behavioral change brought about by AD is the most difficult aspect of the disease to deal with. In the early stages, patients often exhibit tenseness and anxiety, irritability and emotional lability, and depression. They may have difficulty sleeping. They may be very restless and exhibit behaviors such as pacing or tearing paper. They may have hallucinations (believing in the presence of things that are not present) or delusions (belief in things that are not true).

Often, abnormal behavior is triggered by some change in the patient's environment. Examples are a change in location or caregiver, unexpectedly being asked to do something, even if it is not new, like take a bath, or a misunderstood situation perceived as a threat. If such a triggering event can be identified, it can be avoided, thus eliminating or mitigating the behavior.

The Alzheimer's Association in their brochure on Treatments for Behavior, advise that all AD patients who develop behavioral changes should have a thorough medical evaluation, especially if the symptoms appear suddenly. Such an evaluation may reveal a treatable condition contributing to the behavior. Examples they cite include drug side effects, discomfort from infections such as urinary tract, ear, or sinuses, and uncorrected problems from hearing or vision.

The Alzheimer's Association advises that nondrug approaches to correcting or improving behavior should always be tried before turning to prescription drugs. Some of their "coping tips" are as follows:

- Monitor personal comfort, such as pain, hunger, thirst, full bladder, room temperature, and skin irritation.
- Avoid confrontations and arguments.
- Redirect the person's attention if feasible.
- Create a calm environment.
- Try to assure adequate rest.
- Respond to requests.
- Look for triggers to a behavior, and avoid exposing the patient to the trigger.
- Be creative in exploring solutions.
- Don't take the behavior personally.

If nondrug methods do not work or are inadequate, various medications may be helpful. These will be discussed further in the following section.

DRUGS THAT MAY BE HELPFUL FOR BEHAVIORAL SYMPTOMS

Drug approaches to controlling behavioral problems in AD patients should be considered if the symptoms are severe, or if there is a risk the patient may harm themselves or others. The drugs listed here are prescription drugs, and it should be understood that none are US FDA approved to treat behavioral problems in AD patients. This is an "off label" use, where the physician is allowed to prescribe a drug for a different purpose than the condition for which it is approved, if, in the physician's judgment, the drug will be helpful to the patient.

The risks (side effects) should be weighed against the potential benefits and discussed with the patient's family. The initial dose should be low and increased, if necessary, slowly, while carefully monitoring for side effects.

Some of the drugs used for different behavioral problems in AD are as follows:

- Antidepressants are used for mood problems and irritability. Examples of such drugs, among many antidepressant drugs, are Prozac, Paxil, and Zoloft.
- Anxiolytics are used for anxiety, disruptive behavior, and resistance. Examples are Ativan and Serax.

- Antipsychotics are used for hostile aggression, delusions, and hallucinations. There is an FDA-mandated "black box" warning on the label of such drugs because they are associated with an increased risk of stroke and death. Examples of such drugs are Haldol, Seroquel, Risperdal, and Abilify.

EFFORTS TO DEVELOP EFFECTIVE TREATMENTS FOR ALZHEIMER'S DISEASE

As mentioned earlier, there has been considerable effort to develop an effective treatment for AD, but so far nothing has worked in terms of being a really effective treatment. So as said earlier, the "treatment cupboard" for AD is empty except for drugs which give small and temporary improvement. In this section, the amount of effort and its futility so far will be briefly reviewed.

In 2014 Cummings[1] published a review of a decade of AD trials, from 2002 to 2012, using data from Clinicaltrials.gov, a public website that records ongoing clinical trials. During the decade studied, there were 413 AD clinical trials, 124 phase 1, 206 phase 2, and 83 phase 3. These phases are used by the FDA and are related to the status of the drug with respect to treatment development, with phase 1 being early, phase 2 intermediate, and phase 3 near final development. If phase 1 results are positive, then a drug will graduate to phase 2, and if still positive, to phase 3. Phase 3 trials, if adequately successful, are the final trials before the drug is submitted to the FDA for approval.

It was found during this decade that the pharmaceutical industry sponsored 78% of the trials, the National Institutes of Health (NIH) sponsored 7%, 2% were sponsored by a combination of industry and NIH, and organizations such as academic medical centers sponsored 13%. The mechanism of action proposed in the clinical trials was almost 42% for symptomatic treatment, 35% for a disease-modifying molecule, 18% for immunotherapy, about 4% for a device, and about 1% for stem cell therapy. By way of explanation of the immunotherapy entry, considerable attention has been paid to developing antibodies to Aβ, the precursor to amyloid plaques, the concept being that if an antibody binds to Aβ, it might hinder plaque development. So far this approach has not worked.

Cummings[1] looked at the progression of drugs through the three phases: 28% of drugs registered in phase 1 advanced to phase 2 and 8% of drugs registered in phase 2 advanced to phase 3. Only one drug, memantine, was advanced from phase 3 to FDA approval. Memantine is the NMDA antagonist listed in Table 13.1 as giving relatively short-term improvement. This one approval provides a 0.4% overall success rate with a 99.6% failure rate.

Cummings[1] concludes that with AD clinical trials "the number of compounds progressing to regulatory review is amongst the lowest found in any therapeutic area." He also concludes "that relatively few clinical trials are undertaken for AD therapeutics considering the magnitude of the problem."

So in summary, despite 413 clinical trials during the decade studied, this number of trials is small compared to other therapeutic areas with a similar patient burden, such as cancer. Cummings[1] mentions that at the time his paper was published (2014), there were 108 clinical trials for AD therapies, while there were 1438 ongoing trials for oncology agents. Cummings[1] further says that the success rate for oncology agents is 19%, "encouraging biotechnology and pharmaceutical companies to invest time, effort, and funds in oncology drug testing."

Thus, the low success rate in AD clinical trials is no doubt a major reason that the number of trials lags behind that for other major diseases. Speculating on the reason for the low success rate in AD, it seems likely that there is inadequate understanding of the disease pathogenesis in this very complex disease. Not understanding the key steps leading to the disease, means that appropriate therapeutic targets have not been identified.

Since it has been clear for a long time that oxidant damage begins very early in AD pathogenesis, the hypothesis expounded in this book that copper is intimately involved in AD pathogenesis provides new therapeutic targets in AD and hopefully will breathe new life into the search for an effective treatment. Of course, in this book prevention is emphasized by avoiding copper-2 ingestion and decreasing overall copper ingestion. But this line of thinking also opens up the possibility of using anticopper drugs.

Anticopper drugs have primarily been developed for Wilson's disease, an inherited disease of copper accumulation and copper toxicity. One of these drugs, zinc, was developed by the author[2] and approved by the FDA in 1997 for Wilson's disease therapy. Pursuing the concept that anticopper therapy might be helpful in AD, a 6-month double-blind trial of zinc was conducted by the author and colleagues in AD patients. In AD patients 70 years and older, it was shown that cognition decline in zinc-treated patients was significantly stabilized versus patients receiving placebo.[3] The mechanism of the benefit may involve an anticopper effect as originally hypothesized, because the blood-free copper level was significantly decreased by zinc therapy.[3] However, zinc has many beneficial and important effects in the brain. AD patients are zinc deficient, and the AD brain is suspected as being particularly zinc deficient. It has been reported in animals that a genetic defect in the zinc pump which loads neurons with zinc reproduces AD.[4] Thus, zinc therapy may be beneficial in AD patients because of its anticopper effect, or because of restoring important zinc needs in the brain, or both.

In any case, the potential benefit of anticopper therapy, particularly zinc therapy, is the positive note mentioned in the Introduction, when it was stated that "there is possibly some hope that something may be available sooner rather than later."

SUMMARY AND CONCLUSIONS

Reviewing what has been presented here, it is clear that the prevention measures discussed in Chapter 11 become the only way to quell this horrible disease in the foreseeable future. The lifestyle changes revealed to be helpful in Chapter 5, such as exercise, mind games, and ingestion of certain nutrients, and drugs like cholinesterase inhibitors discussed in this chapter help temporarily, particularly in the early stages, to improve memory and cognition. As also discussed in this chapter, behavioral symptoms can often be improved by nondrug approaches such as environmental manipulation. In severe cases, prescription drugs approved for other situations can be used "off label" to help with depression, anxiety, and psychotic behavior.

But all of these approaches are temporary aids, or only treat the symptoms of the disease, not its true cause. Thus, the truth is conveyed in the statement of the Alzheimer's Association quoted earlier, AD is the only disease in the top 10 causes of death that does not have a treatment that will cure, or even slow the progression of, the disease.

Thus, prevention, if a method is available, becomes very important. Here, a hypothesis of causation of the AD epidemic, involving copper-2 ingestion and an increase body loading of copper from increased meat eating, is presented. Since the AD epidemic is causing about 95% of the current cases of AD in developed countries, taking prevention measures covered in Chapter 11 has the potential to prevent 95% of AD. Whether the copper hypothesis is right or not will have to be judged by the reader. There is a large amount of evidence supporting the hypothesis, so it has an excellent chance to be correct. Thus, those who act now on eliminating copper-2 ingestion and limiting copper intake by reducing meat eating will have benefitted greatly when the hypothesis is shown to be correct.

REFERENCES

1. Cummings JL, Morstorf T, Zhong K. Alzheimer's disease drug-development pipeline: few candidates, frequent failures. *Alzheimer's Res Ther* 2014;**6**:37.
2. Brewer GJ, Dick RD, Johnson VD, Brunberg JA, Kluin KJ, Fink JK. The treatment of Wilson's disease with zinc: XV long-term follow-up studies. *J Lab Clin Med* 1998;**132**:264–78.
3. Brewer GJ. Copper excess, zinc deficiency, and cognition loss in Alzheimer's disease. *Biofactors* 2012;**38**(2):107–13.
4. Adlard PA, Parncutt JM, Finkelstein DI, Bush AI. Cognitive loss in zinc transporter-3 knock-out mice: a phenocopy for the synaptic and memory deficits of Alzheimer's disease? *J Neurosci* 2010;**30**:1631–6.

CHAPTER FOURTEEN

Summary and Conclusions

The logical sequence in this book is straightforward. After introducing AD in the first two chapters, in Chapter 3, three very important historical and demographic facts were elaborated and documented as correct. These facts are

1. AD is currently at epidemic prevalence in developed countries.
2. AD prevalence remains low, at about 1%, in undeveloped countries.
3. AD prevalence was low in developed countries before 1900.

Two explanations that are often used to explain these facts were discussed and refuted. The first of these is that since AD is a disease of aging, there were not enough elderly people in developed countries before 1900 to develop AD at a significant prevalence. This was refuted by showing there were plenty of elderly people in the early 1900s in both France and the United States, but they were not getting AD at a significant prevalence. The second explanation is that AD was just attributed to normal aging before 1900, and it was not realized it was a disease entity. This was refuted by demonstrating that textbooks of pathology of that time were not showing the amyloid plaques and neurofibrillary tangles that are hallmarks of AD brain pathology.

This leaves only one viable remaining explanation for the current AD epidemic and that is, environmental change occurring in developed countries since 1900 as the cause. This is such a logical conclusion that it is stated here that these facts cry out, even shout out, "Something in the environment of developed countries introduced in the last century is causing the AD epidemic." Since a minimum of 95% of AD cases in developed countries are related to the epidemic, a very strong case is made that no stone should be left unturned in searching for the environmental cause or causes of the epidemic and trying to mitigate or eliminate them, in order to spare millions of people and their loved ones from the ravages of this horrible disease.

With that as a rationale, Chapters 4 and 5 do a thorough review searching for the causes of the epidemic. To be considered as a possible cause, an agent had to pass two tests. First, exposure to the agent must be greatly increased in the 20th century versus the 19th century, but in developed

Environmental Causes and Prevention Measures for Alzheimer's Disease
ISBN 978-0-12-811162-8
http://dx.doi.org/10.1016/B978-0-12-811162-8.00014-7

countries only, not in undeveloped countries. Second, the agent must have been shown to fit into AD pathogenesis in some manner. In Chapter 4, six metals, aluminum, lead, mercury, zinc, iron, and copper are each considered for possibly being causative of the epidemic. The first three of these are not required for life, they are simply pollutants that are taken up by the body. All three are toxic to the nervous system, and there has been increased exposure to all three in industrialized societies, meaning developed countries. But all three fail the second test, that is they cannot be tied to AD pathogenesis, and they do not produce the amyloid plaques and neurofibrillary tangles characteristic of AD brain pathology. The one possible exception is lead, where monkeys exposed to lead in the first year of life had amyloid plaques in the brain at 23 years of age.[1] However, these abnormalities have not been seen in humans exposed to lead. Zinc, iron, and copper are all essential elements. Exposure to all three has increased in the 20th century in developed countries. But it appears AD patients are zinc deficient, particularly in the brain, where amyloid plaques avidly bind zinc, depleting zinc in other parts of the brain. Iron does not cause AD either, although it contributes to AD pathology by binding to amyloid plaques already formed and increasing release of damaging oxidant radicals.

Copper turns out to be different and is indicted as a major cause of the epidemic. Copper appears to be important in two ways. First inorganic copper, which is divalent–copper, or copper-2, seems to be specifically toxic in causing cognition loss and AD. Sparks and Schruers[2] showed that tiny amounts (0.12 ppm) of inorganic copper in the drinking water of AD animal models greatly enhanced amyloid plaque development and memory loss. Then Morris et al.[3] found that inorganic copper taken in supplement pills containing copper greatly enhanced cognition loss in those who also ate a high-fat diet. Ceko et al.[4] then published that food copper was primarily monovalent copper, or copper-1. This was surprising since living tissue contains both copper-1 and copper-2 as a redox doublet, which catalyzes many reactions important to life. Apparently at death, or harvest, in the absence of oxygen transport, copper-2 is reduced to copper-1. This has important evolutionary implications. Copper is a potentially toxic element, so evolution has developed steps to keep it safely bound, using many chaperones to move copper from one protein to another. But during evolution, organisms were only exposed to copper-1, the copper in food. So a system for handling copper-1 was evolved, but no system for copper-2. Thus, the intestinal cell has a receptor, Ctr1,[5] which binds copper-1, and then routes it through the blood so that it goes to the liver. Ctr1 cannot bind copper-2, and Brewer's group (Hill et al.[6]) has shown that a good portion of ingested copper-2 appears immediately, within 1–2h, in the blood, too soon to have been processed by the liver. Ingested copper-1 in food would not show up in the blood for one or two days, and then, secreted by the liver, it would be safely bound to proteins. It was not until the 20th century that humans in developed countries became exposed to copper-2 and that exposure comes from copper-2 leached into drinking water from copper plumbing and from ingestion of copper-2-containing supplement pills. In summary, it was concluded that copper-2 ingestion is a specific trigger of AD and a major cause of the AD epidemic. Brewer has expounded on this so-called "Copper-2 Hypothesis" in multiple papers since 2009.[7–20]

Secondly, it was found that increased body loading of copper in general, irrespective of copper valence, is a trigger for AD. This finding is based on the studies of the Squitti group in Italy.[21] They found that there was an increased prevalence of patients carrying a variant ATP7B allele in the AD population. ATP7B is the Wilson's disease causing gene. When patients have two disabled ATP7B genes, they have Wilson's disease, an inherited disease of copper accumulation and copper toxicity. Carriers of one disabled ATP7B gene accumulate a little extra copper in their bodies but not enough to require treatment. Assuming the ATP7B variants found at increased frequency in the AD population cause a small increase in body copper load, it can be postulated that carrying a mildly increased copper load for a lifetime is a risk factor for AD. The increased risk from carrying an ATP7B allele would not be a factor in the AD epidemic because the alleles are not common enough to be a factor. But, they may be a marker indicating that an increased body copper load for a lifetime from any source, such as the diet, could be a risk factor for AD. This propels increased copper intake from increased meat eating to the fore. Copper is much better absorbed from meat than from vegetable foods.[22] Meat eating in developed countries has increased by about 35% in the 20th century compared to undeveloped countries.[23] Ingestion of animal fat, which increases toxicity of copper, has increased by a comparable amount. In summary, it was concluded that increased body copper load due to increased meat eating and increased fat ingestion, are contributing significantly to the AD epidemic.

In Chapter 5, the search for environmental factors causative of the AD epidemic continued by looking at diet and other lifestyle factors. This search was stimulated by the many published reports that lifestyle factors can delay the onset of AD or reduce its severity a little once it develops. A good example is increased physical exercise, which appears to do both.

Many lifestyle factors, such as ingestion of nutrients listed on Table 5.1, and physical exercise have minor effects on the disease, or slow its onset, but it is concluded that none other than increased meat and fat ingestion, as already identified in Chapter 4, are causative of the AD epidemic.

Chapters 6 and 7 elaborate further on copper and copper-2, providing additional important information on them. Chapter 8 is important because it pulls all the information together that builds the case that copper-2 is a major causal factor in the AD epidemic—called a "Web of Evidence." The point is made that if you cannot ethically give an agent to humans to prove that the agent is causing the disease, the next best thing to do is develop a "web of evidence" around the claim of causality, so strong that any reasonable person would agree with the assertion of causality. That is what is done in Chapter 8 for copper-2 causality of AD in the AD epidemic. The author believes this "web of evidence" put together in Chapter 8 is so strong that it is close to proof of copper-2 causality.

In Chapter 9, the evidence indicating that an increased body load of copper, irrespective of valence, is also a risk factor for AD, and probably also partially causal of the AD epidemic. Copper is much better absorbed from meat than from vegetable foods,[22] and as a result of a very significant increase in meat eating in developed countries in the 19th century and on,[23] the body load of copper has probably been increased somewhat in people in developed countries during that period, thus contributing to the epidemic.

Thus, we are left with an increase in copper-2 ingestion and an increase in body copper load from increased meat eating as the environmental factors identified here, as causative of the AD epidemic in developed countries in the 20th century and beyond. In Chapter 10, it was shown how these two "AD-triggering" agents fit nicely with known AD risk factors and the two theories of primary AD causation, the amyloid cascade theory and the primary oxidant damage theory.

Having identified the two likely causative factors for the AD epidemic in the earlier parts of the book, it is logical that the next chapter, Chapter 11, explains how to eliminate copper-2 exposure and how to minimize body loading of copper by reducing meat eating. Avoiding copper-2 exposure is relatively easy, since the two main sources of copper-2 ingestion are supplement pills containing copper and ingestion of drinking water containing copper leached from copper plumbing. Supplement pills containing copper can simply be avoided. If drinking water contains copper above 0.01 ppm, a relatively inexpensive device, such as a reverse osmosis device, can be put on the tap furnishing drinking and cooking water, to remove most of the copper.

If one's blood free copper level is above the mean for the population, often about 10 μg/dL, it is probably a good idea to reduce meat eating. A good target is 50% reduction, since a large study has shown that that level of reduction in meat eating will greatly reduce mortality, irrespective of preventing AD.[24]

Chapter 12 discusses US government failures in terms of regulation to lower the risk of AD and thereby help in the fight to eliminate or reduce AD. The two failures are, first that of the US Food and Drug Administration (FDA) not forcing supplement pill makers to eliminate copper-2 from the multimineral supplement pills intended for the general population. There have been two studies now (Morris et al.[3] and Mursa et al.[25]), which show that supplement pills-containing copper-2 are harmful, and it is the FDA's responsibility to protect the public from toxicity from such pills. Second, the US Environmental Protection Agency, the EPA, has the responsibility to assure the safety of drinking water by setting "do not exceed" standards for pollutants in drinking water. The EPA limit for copper in drinking water is 1.3 ppm. Animal studies as far back as 2003[2,26,27] have repeatedly shown that levels 10% that high (0.12 ppm) cause AD in AD animal models. These are two very significant failures of US agencies.

Chapter 13 deals with treatments for AD. The main thrust of this book is on prevention of AD, but these prevention measures will probably not help those already affected with AD. However, the situation is that current treatments are very inadequate. The Alzheimer's Association, as one of its facts about AD, says it is the only disease in the top 10 causes of death that does not have a treatment that will cure, or even significantly slow, the progress of the disease. The FDA has approved five drugs (Table 13.1), but they all help only a little, and for a limited time. There have been significant efforts by pharmaceutical companies, but the failure rate is very high, higher than for any other major disease.

The lack of effective treatment makes the prevention measures discussed in this book all the more important. Simply put, you better keep yourself from getting AD, because once you get it, there are no effective treatments for it, and it is one horrible disease.

So, this is the end of the story, except to discuss the options for you, the reader, in terms of acting on the principle theme of this book, mainly preventing yourself and loved ones from getting AD by avoiding copper-2 ingestion and reducing meat eating. First, it has to be admitted that causation of the AD epidemic by copper-2 ingestion and high meat consumption is still an unproven hypothesis. It is a strongly supported hypothesis, as detailed in Chapters 8 and 9. The best animal studies that could possibly be done to evaluate copper-2 as causative of AD have already been done (Sparks and Schreurs[2]; Sparks et al.[26]; Singh et al.[27]). While it is unethical to give copper-2 to humans to evaluate AD causality, humans have already done it to themselves (Morris et al.[3]) and the results were positive. One can envision intervention studies where a large sample of people are allowed to continue ingesting copper-2 at their normal rate, while another sample of people eliminate copper-2 ingestion, and the rates of development of AD compared. But this kind of study would take a long time (10–20 years?) and be expensive. And there is no commercial interest in such a study, because there is no product. So it would have to be a government-funded study.

A study such as the one proposed above, to add a more definitive proof to the copper-2 part of the hypothesis, is not likely to happen in the foreseeable future. So it is left to each individual reading this book to decide for themselves whether to accept the hypothesis as very likely correct, and then act accordingly and advise their loved ones to act accordingly.

It can be pointed out that there are analogies in this situation to the situation where cigarette smoking was first hypothesized as causing lung cancer and cardiovascular diseases. Those who acted to stop, or never start, smoking at that point had already greatly benefitted when it was later accepted as fact that smoking was indeed causing these diseases. And it can be pointed out that a definitive type of study to prove smoking caused these diseases, by intervening and studying the two groups, was never done with smoking. The data just accumulated until it was clear smoking was causative. It is predicted the same course will occur here. The data will accumulate until the hypothesis is accepted as fact. In the meantime, those who acted preventively based on the likelihood of the hypothesis being valid will have benefitted greatly.

REFERENCES

1. Wu J, Basha MR, Brock B, et al. Alzheimer's disease (AD)-like pathology in aged monkeys after infantile exposure to environmental metal lead (Pb): evidence for a developmental origin and environmental link for AD. *J Neurosci* 2008;**18**:3–9.
2. Sparks DL, Schreurs BG. Trace amounts of copper in water induce beta-amyloid plaques and learning deficits in a rabbit model of Alzheimer's disease. *Proc Natl Acad Sci USA* 2003;**100**:11065–9.
3. Morris MC, Evans DA, Tangney CC, Bienias JL. Dietary copper and high saturated and trans fat intakes associated with cognitive decline. *Arch Neurol* 2006;**63**:1085–8.
4. Ceko MJ, Aitken JB, Harris HH. Speciation of copper in a range of food types by x-ray absorption spectroscopy. *Food Chem* 2014;**164**:50–4.

5. Ohink H, Thiele DJ. How copper traverses cellular membranes through the copper transporter 1, Ctrl. *Ann NY Acad Sci* 2014;**1314**:32–41.
6. Hill GM, Brewer GJ, Juni JE, Prasad AS. Treatment of Wilson's disease with zinc. II. Validation of oral 64 copper with copper balance. *Am J Med Sci* 1986;**292**:344–9.
7. Brewer GJ. The risks of copper toxicity contributing to cognitive decline in the aging population and to Alzheimer's disease. *J Am Coll Nutr* 2009;**28**:238–42.
8. Brewer GJ. The risks of copper and iron toxicity during aging in humans. *Chem Res Toxicol* 2010;**23**(2):319–26.
9. Brewer GJ. Copper toxicity in the general population (editorial). *Clin Neurophysiol* 2010;**121**(4):173–9.
10. Brewer GJ. Toxicity of copper in drinking water (letter to the editor). *J Toxicol Environ Health B Crit Rev* 2010;**13**(6):449–52.
11. Brewer GJ. Issues raised involving the copper hypotheses in the causation of Alzheimer's disease. *Int J Alzheimers Dis* 2011;**2011**:537528. 1–11.
12. Brewer GJ. Copper excess, zinc deficiency, and cognition loss in Alzheimer's disease. *Biofactors* 2012;**38**(2):107–13.
13. Brewer GJ. Copper toxicity in Alzheimer's disease. Cognitive loss from ingestion of inorganic copper. *J Trace Elem Med Biol* 2012;**26**:89–92.
14. Brewer GJ. Metals in the causation and treatment of Wilson's disease and Alzheimer's disease and copper lowering therapy in medicine. *Inorg Chim Acta* 2012;**393**:135–41.
15. Brewer GJ, Kaur S. Zinc deficiency and zinc therapy efficacy with reduction of serum free copper in Alzheimer's disease. *Int J Alzheimers Dis* 2013:586365. 1–4.
16. Brewer GJ. Alzheimer's disease causation by copper toxicity and treatment with zinc. *Front Aging Neurosci* May 2014;**16**(6):92.
17. Brewer GJ, Kaur S. Ingestion of inorganic copper from drinking water and supplements is a major factor in the epidemic of Alzheimer's disease. In: Rothkopf MM, editor. *Metabolic medicine and surgery*. Boca Raton (FL): CRC Press; 2014.
18. Brewer GJ. Divalent copper as a major triggering agent in Alzheimer's disease. *J Alzheimers Dis* 2015;**46**:593–604.
19. Brewer GJ. Copper-2 ingestion, plus increased meat eating leading to increased copper absorption, one major factor behind the current epidemic of Alzheimer's disease. *Nutrients* 2015;**7**:10053–64.
20. Brewer GJ. Copper-2 hypothesis for causation of the current Alzheimer's disease epidemic together with dietary changes that enhance the epidemic. *Chem Res Toxicol* 2017;**30**:763–8.
21. Squitti R, Polimanti R, Siotto M, et al. ATP7B variants as modulators of copper dyshomeostasis in Alzheimer's disease. *Neuromol Med* 2013;**15**:515–22.
22. Brewer GJ, Yuzbasiyan-Gurkan V, Dick R, Wang Y, Johnson V. Does a vegetarian diet control Wilson's disease? *J Am Coll Nutr* 1993;**12**:527–30.
23. Teicholz N. How Americans got red meat wrong. *Atlantic* June 2, 2014.
24. Sinha R, Gross AJ, Graubard BI, Leitzmann MF, Schotzkin A. Meat intake and mortality: a prospective study of over half a million people. *Arch Int Med* 2009;**167**:562–71.
25. Mursa J, Robein K, Hamack LJ, Park K, Jacobs DR. Dietary supplements and mortality rate in older women: the Iowa Women's Health Study. *Arch Int Med* 2011;**445**:1625–33.
26. Sparks DL, Friedland R, Petanceska S, et al. Trace copper levels in drinking water, but not zinc or aluminum, influence CNS Alzheimer-like pathology. *J Nutr Health Aging* 2006;**10**:247–54.
27. Singh I, Sagare AP, Coma M, Perlmutter D. Low levels of copper disrupt brain amyloid-beta homeostasis by altering its production and clearance. *Proc Natl Acad Sci USA* 2013;**110**:14471–6.

INDEX

Printed in the United States
By Bookmasters